Contents

The Wild Garden

The Wild Garden

An illustrated guide to weeds

Lys de Bray

Weidenfeld and Nicolson
London

Designed by Juanita Grout
for George Weidenfeld and Nicolson Limited
11 St John's Hill, London SW11

ISBN 0 297 77480 8

Colour separations by Newsele Litho Ltd
Printed in Great Britain by
Cox and Wyman Ltd
London, Fakenham and Reading

Preface

Weeds and wild flowers are distinguished only by their habitat. When buttercups shine golden in a meadow, we love them, but if they should come up in the rose-bed or among the onions, we call them weeds because they grow there uninvited, unwanted and unplanted.

Some of the oldest cottage-garden flowers, such as Columbine, Marigold and Forget-me-not glorify the cracks in our paths and shame us because there never seems time to 'weed' them. But does it matter? These and other old favourites have such a history as medicinal plants, vegetables, garnishes and housewives' aids that they have earned our respect and affection and perhaps deserve to stay. They will often bloom, sturdy and indomitable, in the poor soil left when the builders depart.

Other garden visitors began life as delicate and pampered hot-house dwellers, which escaped and colonized happily. Some 'garden' flowers and even shrubs and trees may be classed as weeds because they grow too vigorously and multiply too rapidly in the nurtured soil of our gardens.

There is, indeed, a rich and historically interesting heritage of uninvited guests as we walk from the back door to the bonfire; some of them are edible, some will heal us, some are aesthetically beautiful, some scent the air around and one or two are dangerous, while others have fascinating medical, religious, magical and historical connotations that they have gathered about them over the centuries.

The Wild Garden touches on all these aspects of the most common of the plants that we know as 'weeds'. It is not intended to be comprehensive and the selection of plants is inevitably personal. My aim has been to give pleasure and practical help to the book's readers, whether they use it as an aid to recognition, for horticultural help, as a therapeutic compendium, as a back-to-the-land recipe collection, or as a guide to the total annihilation of their personal weed population.

The book has been arranged in sections according to flower-colour in order to help identification. (The few non-flowering plants, such as ferns, all fall within the green section.) Within these colour groupings, the plants are placed in their taxonomic order according to the *Flora of the British Isles* by Clapham, Tutin and Warburg, Cambridge University Press, to whom I am greatly indebted for permission to use the abbreviated botanical descriptions in this book. The common

names of the plant, with which most people will be more
familiar, are listed first and followed by the botanical name.
The glossary on page 11 lists the terms used throughout
the text and includes outline drawings of the main flower- and
leaf-shapes.

Acknowledgments

My most grateful thanks are due to my husband, Larry Blonstein, who encouraged, listened, criticized constructively, cooked life-saving meals and worked on the chemical section; to my secretary Carole who burnt out the typewriter in the week before the text deadline; to Peter Hatherley for checking the medical information; to Michael Belkin, Ph.c, MPS for translating medieval measures into present-day understanding; to Pat Halliday at Kew who assisted with taxonomy and enthusiasm; to Janet Augustin who taught me some spells, and to Andrew J. Marsh, ND, DO, D.Hom, who checked the natural therapy. Last, but most important, my gratitude for the harvest of weeds, so generously hurled over my fence, deposited in baskets and bags and barrows on back step and at front door and passed through kitchen windows by kind and thoughtful gardening friends polite enough to refrain from comment, at least until out of earshot.

Glossary of Terms

ACUMINATE

AWN

AXILLARY

BIPINNATE

A-BRACTEOLES

CORDATE

CORYMB

CRENATE

ACUMINATE having a gradually diminishing point

ANNUAL a plant completing its life-cycle within twelve months of germination

AWN a stiff bristle-like projection from the tip or back of the lemma in grasses

AXIL the angle between the stem and the leaf stalk

AXILLARY arising in the axil of a leaf or bract

BIENNIAL completing its life-cycle within two years and not flowering in the first year

BIFID split deeply in two

BIPINNATE (2-pinnate) a leaf in which the primary divisions are themselves pinnate, that is, a leaf composed of more than three leaflets, arranged in two rows along a common stalk

BRACT modified leaf at the base of a flower

BRACTEOLES minute bracts

CALCAREOUS containing lime

CALCIFUGE a plant characteristic of limy or basic soil

CAMPANULATE bell-shaped

CILIATE with regularly arranged hairs projecting from the leaf-margin

CONNATE organs of the same kind growing together and becoming joined, though distinct in origin

CORDATE heart-shaped

CORM a short, swollen, bulb-like underground stem

CORYMB a raceme with the pedicels becoming shorter towards the top, so that all the flowers are at the same level

CORYMBOSE growing in corymbs

CRENATE with rounded marginal teeth

CUNEATE wedge-shaped

CUNEIFORM wedge-shaped with

CUNEATE

CYME

DELTOID

ELLIPTIC

HASTATE

LANCEOLATE

LOBE

the thin end at the base

CUSPIDATE spear-shaped at the tip

CYME a flower-head with a broad and flattened appearance

DECUMBENT stems lying on the ground and tending to rise at the growing tip

DELTOID shaped like a triangle

DENTATE toothed

DIFFUSE loosely spreading

DIOECIOUS having the male and female flowers on separate plants

EFFUSE spreading widely

ELLIPTIC oval with pointed ends

ENSIFORM long and narrow but wider at the centre

FIBROUS formed of fibres

GLABRESCENT becoming glabrous

GLABROUS without hairs

GLAUCOUS bluish

GLUME a small bract with an axillary flower (as in most grasses)

HASTATE spear-shaped

HEMISPHERICAL half a sphere

HISPID coarsely and stiffly hairy

IMBRICATE of organs with their edges overlapping, when in bud, like the tiles on a roof

INFLORESCENCE the flowering branch, or portion of the stem above the last stem leaves, including its branches, bracts and flowers

KEEL a sharp edge resembling the keel of a boat; the lower petal or petals when so shaped

LACINIATE deeply and irregularly divided into narrow segments

LANCEOLATE narrow, tapering at each end

LATERAL attached to, or near, the sides of an organ

LAX loose

LEMMA the outer bract surrounding the grass flower

LINEAR slender, with parallel sides

LOBE the divided edge of a leaf (but

MUCRONATE

not into separated leaflets)

LYRATE shaped like a lyre

MUCRONATE, MUCRONULATE with a short, narrow point

NODE a point on the stem where one or more leaves arise

OBLANCEOLATE opposite or inversely

OBOVATE broadest above the middle

OBTUSE

OBTUSE blunt

ORBICULAR rounded

OVATE egg-shaped

OVATE-ACUTE egg-shaped with a point

ORBICULAR

OVOID of a solid object which is ovate in longitudinal section

PALMATE consisting of more than three leaflets arising from the same point

PANICLE a branched inflorescence; more strictly, a branched racemose inflorescence

OVATE

PARTITE parted or divided to, or almost to, the base

PECTINATE lobed, but with the lobes resembling the teeth of a comb

PEDICEL the stalk of a single flower

PELTATE of a flat leaf with its stalk on the centre of the under surface

PALMATE

PENTAGONAL pertaining to a five-sided or five-cornered figure

PERENNIAL a plant living for more than two years and usually flowering each year

PETIOLE the stalk of a leaf

PILOSE hairy, with rather long, soft hairs

PINNA each primary division of a pinnate or pinnatifid leaf, especially in ferns

PINNATE a leaf with leaflets arranged on opposite sides of a common stalk or rhachis

PINNATE

PINNATIFID pinnately cut, but not into separate portions

PINNATISECT similar to pinnatifid but with some of the lower divisions cut to the midrib

POLYGONAL many-sided

PROCUMBENT lying loosely along the surface of the ground

PUBESCENT shortly and softly hairy

RACEME an unbranched inflorescence where the flowers are borne on pedicels

RACEMOSE a (usually) conically shaped inflorescence, whose growing

points add to the inflorescence. The youngest and smallest branches or flowers are normally nearest the apex

RENIFORM

RENIFORM kidney-shaped

RHIZOME an underground stem lasting more than one growing season

ROSETTE a cluster of leaves growing from a central point, usually flat and circular

RUNCINATE pinnately lobed with the leaves directed backwards, towards the leaf base

SAGITTATE

SAGITTATE shaped like an arrowhead

SCABRID rough to the touch

SEGMENT a division of a plant organ (leaf, calyx, corolla, etc.), which may be cleft

SERRATE

SERRATE toothed like a saw

SESSILE without a stalk

SINUATE having a wavy outline

SPADIX a fleshy spike, often terminating in a club-shaped tip

SPADIX

SPATHE a large bract which encloses the flower head or spadix

SPATHULATE paddle-shaped

SPIKELET a small spike, as in the inflorescence of grasses

STOLON a creeping stem produced from the central part of a plant, rooting at intervals and usually above ground

SPATHE

SUB-CAMPANULATE in combinations e.g. not quite, nearly, campanulate

SUBULATE awl-shaped

SUCCULENT fleshy, juicy

TAP-ROOT a long, tapering cylindrical root

SPATHULATE

TENDRILS a climbing organ formed of the whole or part of the stem, leaf or petiole

TERETE not ridged, grooved or angled

TRIFOLIATE having three leaflets

TRIGONOUS of a solid body triangular in section but obtusely angled

SUBULATE

TRIQUETROUS of a solid body triangular in section and acutely angled

TRISECT cut into three almost separate parts

TRUNCATE cut off abruptly; cut off at the end

TAP-ROOT

TUBER a swollen part of an underground stem or root, of one

year's duration

UMBEL an inflorescence in which the pedicels all arise from the top of the main stem. An umbrella-shaped inflorescence

UNDULATE wavy

URCEOLATE

URCEOLATE globular to subcylindrical but strongly contracted at the mouth

WHORL a ring of flowers and/or leaves all arising from the stem at the same level as each other

WING the lateral petals in the flowers of Papilionaceae and Fumariaceae

Conversion Tables

LENGTH

Imperial to Metric
1 in = 2.54 cm (25.4 mm)
1 ft = 30.5 cm
1 yd = 0.91 m

Metric to Imperial
1 mm = 0.04 in
1 cm = 0.4 in
1 m = 39.4 in

AREA

Imperial to Metric
1 sq in = 6.5 sq cm
1 sq ft = 929 sq cm
1 sq yd = 0.84 sq m
1 acre = 0.4 ha

Metric to Imperial
1 sq cm = 0.16 sq in
1 sq m = 1.2 sq yd
1 ha = 2.47 acres

LIQUID VOLUME

Imperial to Metric
1 fl oz = 28.4 ml
1 gill = 142 ml
1 pint = 0.57 l
1 qt = 1.14 l
1 gal = 4.56 l

Metric to Imperial
1 ml = 0.03 fl oz
1 l = 0.22 gal

Household to Metric
1 teaspoon = 5 ml
1 dessertspoon = 10 ml
1 tablespoon = 25 ml
1 eggcup = 35 ml
1 wineglass = 150 ml
1 breakfast cup = 250 ml

Household to Imperial
1 teaspoon = about $\frac{1}{8}$ fl oz
1 dessertspoon = about $\frac{1}{3}$ fl oz
1 tablespoon = about $\frac{3}{4}$ fl oz
1 eggcup = about $1\frac{1}{4}$ fl oz
1 wineglass = about 5 fl oz
1 breakfast cup = about 8 fl oz

WEIGHT

Imperial to Metric
1 oz = 28.4 g
1 lb = 454 g
1 st = 6.4 kg
1 cwt = 51 kg
1 ton = 1017 kg

Metric to Imperial
1 g = 0.035 oz
1 kg = 2.2 lb
1 tonne = 2200 lb

Key to Abbreviations

A   Appearance

AH Average height

L   Leaves

R   Root

F   Flowers

H   Habitat

Yellow-flowered and Orange-flowered species

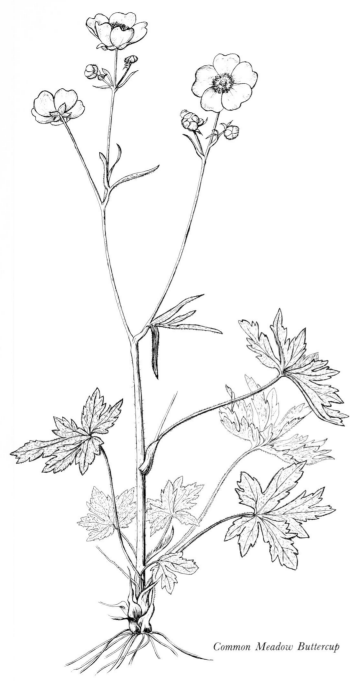

Common Meadow Buttercup

Common Meadow Buttercup, Crowfoot, Gold Cup, Meadowbloom, Yellows, Bachelor's Buttons, Blister Weed

Ranunculus acris L.

A   Much branched, erect
AH 15–100 cm (6–40 in)
L   Palmately 2–7 lobed and segmented, hairy
F   An irregular cyme, glossy, bright yellow,
    saucer-shaped. April to September
H   Grassland
    Perennial
    Poisonous

Buttercups are lovely to look at in the mass but they produce an acrid juice and fatalities have occurred with young animals out grazing for the first time.

The saucer-shaped, shiny surfaces of the petals form reflectors which concentrate solar radiation on to the centre of the flower, causing a slight increase in temperature at the 'focus' of the flower-head which attracts pollinating insects. The roots secrete a nitrogen-poisoning substance which prevents the growth of clovers.

The buttercup family has many members that rightfully dwell in the herbaceous border, such as Larkspur, Delphinium, Peony, Globe Flower and all the varieties of Clematis – these are valued favourites. However, keep the buttercup tribe out of the vegetable garden, because their presence will stunt the growth of peas and beans, and will force strawberry plants into premature growth.

An extract of the fresh plant has been used to alleviate skin problems and to cure arthritis and sciatica but it is dangerous to experiment with this plant – qualified homoeopathic advice should be carefully followed as to procedure and dosage.

CONTROL Hand-weed individual plants, using gloves.

RANUNCULACEAE Buttercup Family

Creeping Meadow Buttercup

Ranunculus repens L.

A   Creeping, with erect flowering stems
AH 15–60 cm (6–24 in)
L   Triangular, 3 lobed, the lobes further cut, hairy
R   Stoloniferous

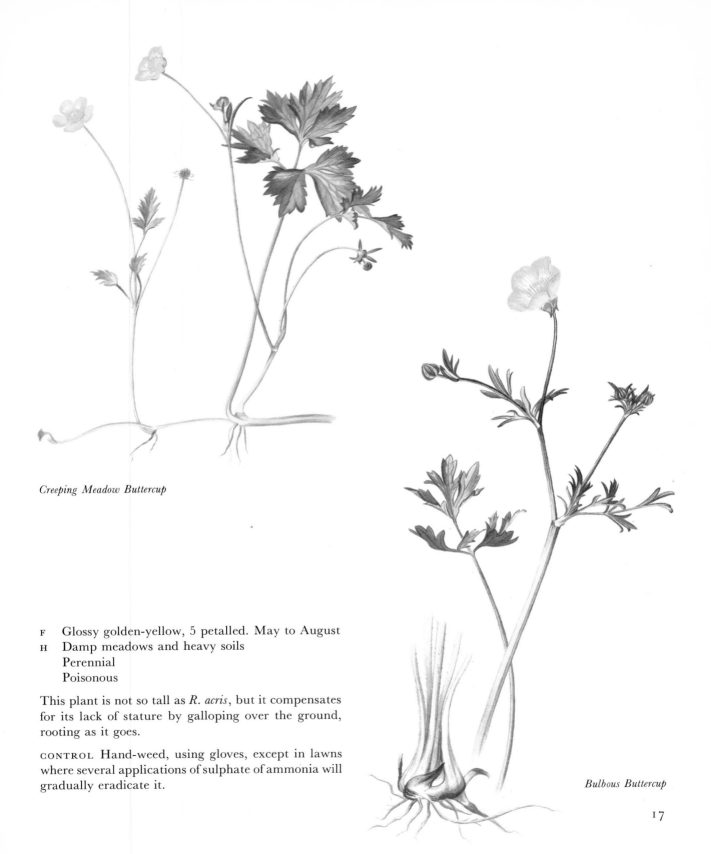

Creeping Meadow Buttercup

F   Glossy golden-yellow, 5 petalled. May to August
H   Damp meadows and heavy soils
     Perennial
     Poisonous

This plant is not so tall as *R. acris*, but it compensates for its lack of stature by galloping over the ground, rooting as it goes.

CONTROL Hand-weed, using gloves, except in lawns where several applications of sulphate of ammonia will gradually eradicate it.

Bulbous Buttercup

17

Bulbous Buttercup, Cuckoo Buds, St Anthony's Turnip, Frogwort
Ranunculus bulbosus L.

A   Erect
AH 15–40 cm (6–16 in)
L   Ovate, 3 lobed, cut into linear segments, hairy
R   Corm-like stem tuber
F   An irregular cyme, glossy, bright yellow. May to June
H   Dry pastures
    Perennial
    Poisonous

This buttercup is very variable in its aerial parts, but is easily distinguished by the stem-tubers or 'bulbs' and the reflexed sepals, and it is earlier flowering than the other two field buttercups (*R. acris* and *R. repens*).

Bulbous Buttercups were used in the Middle Ages to 'blister' victims of the plague in an attempt to draw the poison in the body to the surface of the skin. According to the old herbalist Culpeper, 'Virgins, in ancient time, used to make powder of them to furrow their beds', but sadly he is not specific as to the purpose of the powder.

CONTROL Easily pulled up on a damp day, if they should appear in the vegetable garden.

Celandine, Lesser Celandine, Figwort, Pilewort
Ranunculus ficaria L.

A   Branched, ascending
AH 5–25 cm (2–10 in)
L   Patterned light and dark green, cordate
R   Fusiform tubers
F   8–12 petals, glossy yellow. March to May
H   Deciduous woodland, hedgebanks
    Perennial
    Poisonous

A welcome harbinger of spring, the shining stars of the Lesser Celandine are the first really bright flowers of the year. (It should not be confused with the Greater Celandine, *Chelidonium majus*, which flowers much later and which has a very different habitat.)

USES The Celandine's other name of Pilewort is an earthy indication of the plant's early use as a remedy for haemorrhoids. Bruise the leaves and stems in a mortar with a little pure lard and apply as an ointment or suppository. Culpeper in 1653 said that 'the very herb borne about one's body next the skin helps in such diseases, though it never touches the place grieved' – this method is for the cowardly.

CONTROL Root it out while you can still see the leaves and flowers. Once these have died away, it is difficult to find except by grubbing about for the miniature hanging clumps of tubers, each of which will regenerate next year into another plant. The Celandine multiplies by means of tiny bulbils found in the leaf-axils of unpollinated plants. These bulbils fall to the ground and can be washed some distance away by the heavy spring rains. Each tiny bulbil will grow into a small tuber which in turn will grow into a new plant, though it will not flower in its first year.

Greater Celandine, Swallow-wort, Tetterwort
Chelidonium majus L.

A   Erect and leafy
AH 30–90 cm (12–36 in)
L   Glaucous, slightly hairy, 5–7 ovate leaflets
F   4 petals, yellow, crinkled. May to August
H   Walls and paving, banks and hedgerows
    Perennial
    Poisonous

Despite its name this plant is not related to the Celandine. *Chelidonium* is a poppy and shares the creased-taffeta look about its petals that all poppies have. The name Swallow-wort probably derives from the fact that it flowers when the swallows return to nest in the spring. According to an ancient herbal myth, the parent swallows sought out the plant when their nestlings were blinded by some accident and laid it upon the eyes of their young, whereupon their sight was immediately restored.

The plant is very poisonous indeed and should not be ingested under any circumstances – a couple of ounces of the expressed juice have been known to kill a large dog. However, there is an old charm which is perfectly harmless – 'He who carries it upon his person together with the heart of a mole will vanquish his enemies and win his lawsuits.'

USES Very small amounts of this plant are used by qualified homoeopathic practitioners in remedies for eczema, scrofulous diseases and jaundice. The acrid orange juice which is exuded from any cut part of the plant may be used to cure warts. Break off a leaf and touch *only* the wart with the end of the oozing stem. Do not let the orange juice – a latex – spread on to the surrounding skin, or it may redden and blister. Apply fresh juice daily. This works very well on some people and not at all on others. The plant may be used for corns as well, but *not* for verrucas.

CONTROL This plant looks very attractive growing between cracks in old paving. However, if it must be purged from paths or rockery, wear gloves to pull it up; as well as being dangerous, the juice stains and smells most unpleasant.

FUMARIACEAE Fumitory Family

Yellow Corydalis, Yellow Fumitory
Corydalis lutea L.

A   Weakly branching
AH 15–30 cm (6–12 in)
L   Pinnate
R   Fibrous
F   Tubular, yellow, with a short spur. May onwards
H   Old walls
    Perennial

This lime-loving plant will be found clothing old walls

with attractive tufts of feathery, light-green leaves which are useful foliage for flower arrangers. It will grow in sun or semi-shade.

CONTROL Easily eradicated with a proprietary herbicide.

CRUCIFERAE Cabbage Family

Charlock, Runch, Field Mustard
Sinapis arvensis L.

A   Erect branching stem
AH 30–80 cm (12–32 in)
L   Lyrate, hairy
R   Tap-root
F   4 petals, yellow. May to August
H   Arable land
    Annual

This is the commonest and most troublesome weed of cornfields, and from a distance the expanse of bright yellow flowers could be mistaken for a planned sowing. The seeds which were formerly used in the manufacture of mustard oil, particularly in France, can remain fertile for anything between 10 and 50 years, even under grass.

It is the food of the Orange Tip butterfly, *Anthocharis cardamines* L.

CONTROL It is unlikely to be a pest in a cultivated garden, but to remove from a field or other large area, spray before the plants are 8 cm (3 in) high with either 4 per cent solution of copper sulphate, or a 15 per cent solution of iron sulphate.

Celandine

Greater Celandine

20

Yellow Corydalis

Charlock

21

Common Yellow Rocket

CRUCIFERAE Cabbage Family

Common Yellow Rocket, Winter Cress, Land Cress, Bitter Cress, Herb St Barbara, Water Mustard
Barbarea vulgaris R. Br.

A   Erect branching stem
AH 30–90 cm (12–36 in)
L   Deep green, shining, with cordate terminal lobe and coarsely toothed
R   Yellowish tap-root
F   Bright yellow, 4 petals. May to August
H   Hedges and waysides
    Biennial

The Latin name of this plant commemorates Saint Barbara, because the leaves are still available on her saint's day, 4 December. She is the patron saint of firearms and used to be invoked against the terrors of lightning.

USES A member of the great cabbage family, the buds and leaves are a valuable source of vitamin C and the plant is used in America as a salad vegetable. It is sometimes called Winter Cress because its leaves are available when there is little else: they are rather bitter in flavour.

CONTROL Do not allow to seed, as, like all the *Cruciferae*, its progeny will be legion. It is easily weeded out by hand.

CRUCIFERAE Cabbage Family

Hedge Mustard, Singer's Plant
Sisymbrium officinale L.

A   Branched like a candelabra
AH 30–90 cm (12–36 in)
L   A basal rosette, deeply pinnatifid
F   Corymbose, small-flowered, pale-yellow petals. June to August
H   Hedgebanks, roadsides, waste places
    Annual

Found almost everywhere, this is one of those unmemorable plants – not very tall, not very interesting in shape, with small, rather dull-yellow flowers, seeming always to be furred with dust.

It is the food of the Orange Tip butterfly, *Anthocharis cardamines* L.

USES The plant was used in France up to the time of Louis XIV as an infallible remedy for loss of voice and all diseases of the throat. It is an expectorant, a diuretic and a stomachic, and is used as a tea for congestion, catarrh and laryngitis.

To make an infusion: steep 1 tsp of the plant in $\frac{1}{2}$ cup of water for 4–5 minutes. Dosage: $1\frac{1}{2}$–2 cups a day, a mouthful at a time. For catarrhal problems, sweeten with honey if desired.

CONTROL Hand-weed.

TROPAEOLACEAE Nasturtium Family

Nasturtium, Great Indian Cress
Tropaeolum majus L.

A Trailing
L Orbicular to reniform, peltate, nerved
F Showy, orange, yellow, red, creams or mixtures of these. June until frosts
H A garden escape
   Annual

In its native country of Peru this familiar flower is a perennial, but in northern climates it is reduced to a squashy pulp by the first frost of the autumn. If the strange, helmet-shaped flowers were rare and difficult to grow, they would probably be in great demand; as it is, the Nasturtium is often cursed for its lusty, sprawling growth which, because of good garden soil, is often productive of more leaves than flowers. Gerard had noticed that they 'dispersed themselves far abroad, by means whereof one plant doth occupie a great circuit of ground'.

It is said that woolly aphides and white fly are repelled by the presence of Nasturtiums and experiments are continuing with the symbiotic association of this plant grown at the same time as potatoes, radishes and tomatoes.

Nasturtiums are reputed to have powers of rejuvenation and are also said to have aphrodisiac qualities – another of its names being 'Flower of Love'.

USES Nasturtium leaves, flowers and seeds may all be eaten; the seeds and flower buds may be pickled like capers, while the flowers and young leaves make an attractive garnish in salads.

The plant contains a high proportion of sulphur, and an excellent hair-lotion may be made as follows: take 3 oz (85 g) each of fresh Nasturtiums (leaves, seeds and flowers), Stinging Nettle leaves and Box leaves, and 1 pt (0.6 l) 90 per cent alcohol. Mince the leaves in an ordinary mincer, retaining all juices. Allow to marinade in the alcohol for a fortnight. Strain and add a few drops of oil of Rosemary (or Lemon Verbena). Use often for brisk scalp massage, and sprinkle on the hairbrush before brushing the hair.

An infusion of the leaves ($\frac{1}{4}$ oz (7 g) to $\frac{1}{2}$ pt (0.3 l) boiling water) is a remedy for those who suffer from catarrh and bronchitis. Make fresh every day and take in 3 equal quantities.

CONTROL The ropes of succulent stems and leaves are easily rooted up. Dig up the seedlings as soon as recognized and move them to a hot barren patch of poor soil, if you wish to grow plants with fewer leaves and far more flowers. Nasturtiums contain phosphorus and sulphur and are very good in the compost-heap.

If your Nasturtiums are harbouring black fly, solve two problems at the same time by making a pesticide from collected cigarette dog-ends. Use about 40 to 60 cigarette butts to 1 gal (4.5 l) water, boil for half an hour, and strain. This is a deadly poison, so label it as such and keep it away from children. Dilute when needed at a rate of 4 parts water to 1 part concentrated nicotine solution, and do not spray vegetables that are to be eaten.

OXALIDACEAE Wood-Sorrel Family

Yellow Sorrel, Yellow Oxalis
Oxalis corniculata L.

A   Stems weak, procumbent
AH 5–15 cm (2–6 in)
L   Trifoliate

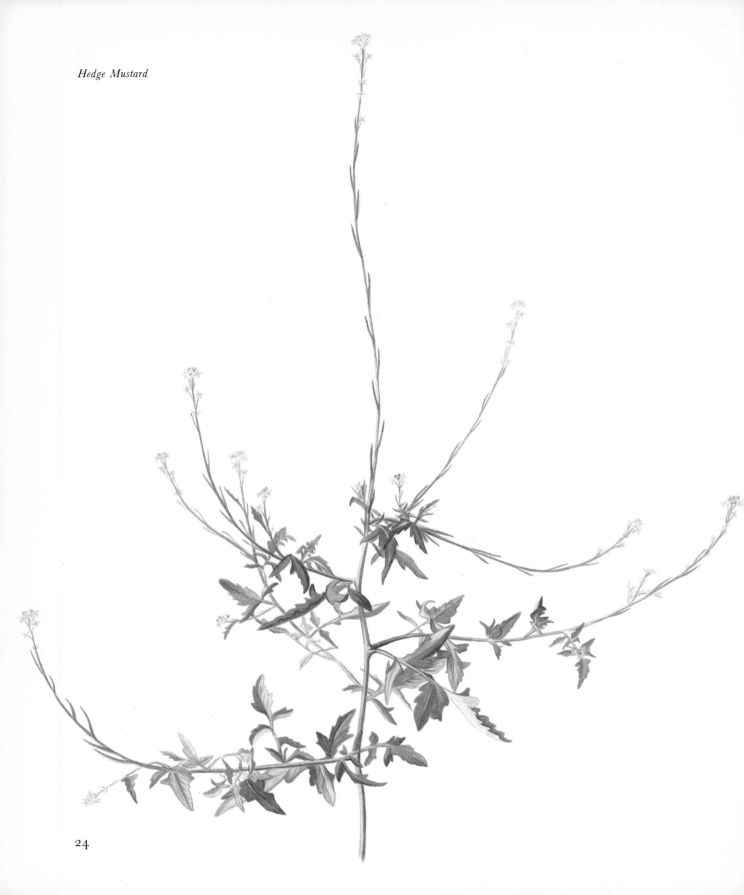

Hedge Mustard

24

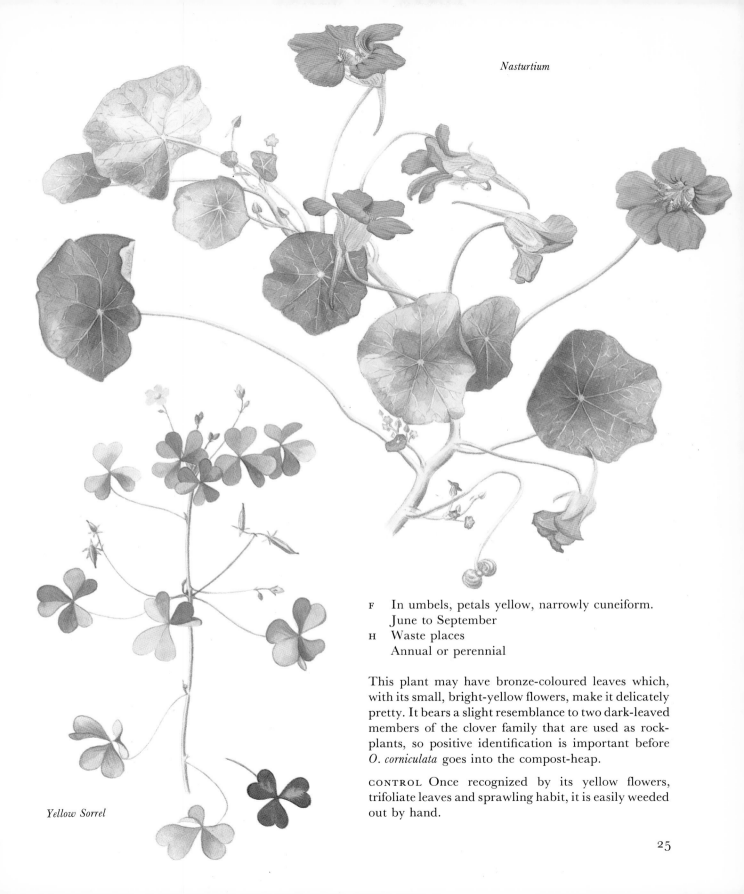

Nasturtium

Yellow Sorrel

F   In umbels, petals yellow, narrowly cuneiform.
     June to September
H   Waste places
     Annual or perennial

This plant may have bronze-coloured leaves which, with its small, bright-yellow flowers, make it delicately pretty. It bears a slight resemblance to two dark-leaved members of the clover family that are used as rock-plants, so positive identification is important before *O. corniculata* goes into the compost-heap.

CONTROL Once recognized by its yellow flowers, trifoliate leaves and sprawling habit, it is easily weeded out by hand.

25

Tree Lupin

PAPILIONACEAE Pea Family

Tree Lupin
Lupinus arboreus Sims

A Bushy shrub
AH Up to 3 m (10 ft)
L Glabrous, oblanceolate, mucronulate
F In racemes, yellow. June to September
H Waste places and sea shores
   Poisonous

A native of California, the Tree Lupin is found in Europe in sandy gardens that are near the sea or back on to a railway line. It is easily recognized by its bright-green, lupin-like leaves, and when in bloom, by the scented, yellow flower-spikes.

This plant thrives on light, poor, sandy soil, where nothing much else will grow, and is much more pleasing to look at than Brambles as a ground-cover plant. In association with heathers it can be most attractive in the wild garden, as the leaves are evergreen and the flowers very colourful.

In the past, Tree Lupins were used in wine- and liqueur-making, and opinion varies as to the toxicity of the plant. The seeds are certainly poisonous if taken in excess, resulting in stomach pains, diarrhoea and vomiting. An emetic should be given if it is known that children have eaten lupin seeds, though they are so bitter that the taste is usually a sufficient deterrent. The seeds contain lupinin.

CONTROL Weed out by hand when the plant is young.

PAPILIONACEAE Pea Family

Gorse, Furze, Whin
Ulex europaeus L.

A Dense shrub
AH 60–200 cm (24–80 in)
L Deeply furrowed spines
F Yellow, wings longer than keel. March to June
H Heaths

'Kissing's out of season when the Gorse is out of bloom' – so runs the old saying. The sweet scent of a hillside of Gorse in blossom is part of summer memories, though solitary sprays of blossom may be found at all seasons. The seed-pods burst with a cracking sound on hot days, flinging the seeds in all directions, a sound which country people described as 'the fairies shooting'. The bushes used to be used as fodder in winter and were bruised and crushed to soften the thorns.

Strangely enough, Gorse is not as hardy as may be imagined and in very cold winters frost will kill it.

It is the food plant of the Green Hairstreak butterfly, *Callophrys rubi* L.

USES The flowers can be used to make perfume and wine. The following is a New Forest recipe for Gorse blossom champagne: take $\frac{1}{2}$ gal (2.25 l) measure of Gorse blossoms, 1 gal (4.5 l) of hot water, 1 lemon, 2 lb (905 g) of cane sugar or $1\frac{1}{2}$ lb (680 g) sugar and $\frac{1}{2}$ lb (225 g) pure honey, 1 lb (455 g) raisins, large if possible, 1 thick slice toasted brown bread and 1 oz (30 g) yeast.

Put blossoms, sugar and hot water into a large enamel bowl and leave for one week. At the end of this period press down blossoms and stir well. Do this each day for a week and then strain out the solids.

Into the strained liquid put the raisins cut small and the lemon thinly sliced. Spread the toast both sides with the yeast, and put this to float on the brew. Leave for 14 days, skim off any floating residue, and strain until the liquid is clear, then bottle.
(N.B. The blossoms must be dry for gathering. Be sure to cork the bottles tightly, and use natural cork stoppers, not plastic ones.)

CONTROL As soon as the tiny seedling plants are recognized (the very smallest do not have spines), they should be removed by hand. Full-grown Gorse is a stout defender of its territory and the strong roots can penetrate into cliffs and stony ground, until fire is the only way to clear the land. Even then the blackened roots will remain and the ground will have to be rotovated to clear it thoroughly.

If you really love the scent of Gorse in summer, a compromise would be to grow the handsome garden variety, *Flore plena*, which has a powerful scent of almond-coconut with a suggestion of orange.

PAPILIONACEAE Pea Family

Black Medick, Blackseed
Medicago lupulina L.

A Procumbent or ascending
AH 5–50 cm (2–20 in)
L Leaflets obovate, serrate

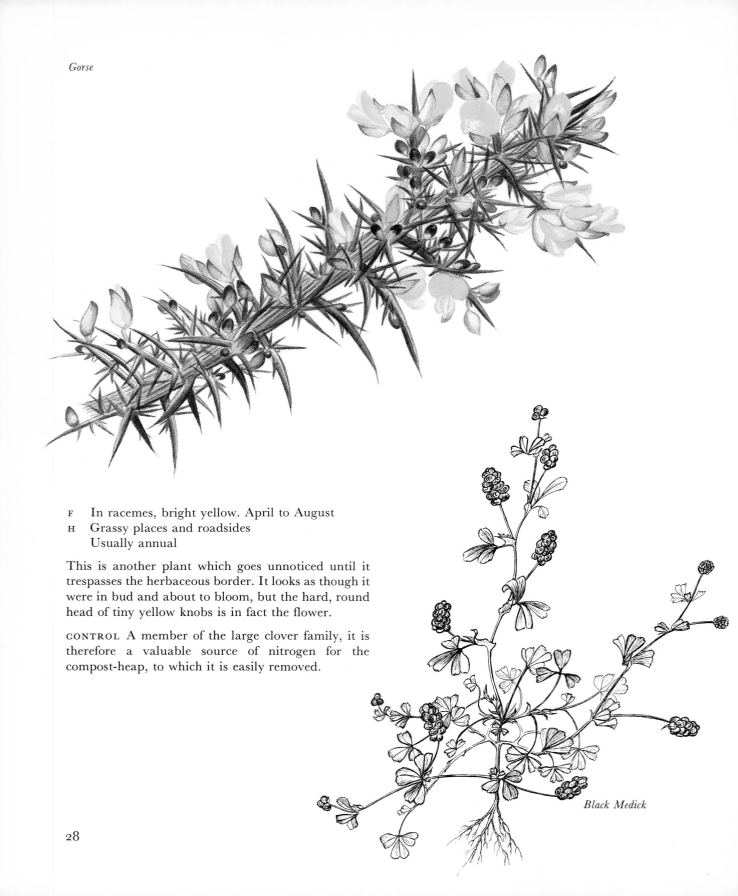

F   In racemes, bright yellow. April to August
H   Grassy places and roadsides
     Usually annual

This is another plant which goes unnoticed until it trespasses the herbaceous border. It looks as though it were in bud and about to bloom, but the hard, round head of tiny yellow knobs is in fact the flower.

CONTROL A member of the large clover family, it is therefore a valuable source of nitrogen for the compost-heap, to which it is easily removed.

Black Medick

28

PAPILIONACEAE Pea Family

Spotted Medick, Calvary Clover, Cogweed
Medicago arabica L. Huds

A Procumbent
AH 10–60 cm (4–24 in)
L Leaflets obovate, serrate, usually blotched
F Racemes 5–7 flowered, flowers yellow. April to
   August
H Grassy places and waste ground
   Annual

Occasionally used in the rock-garden for its interesting leaves, this plant will grow tall and thick in good soil and will quickly smother anything small growing near it. It is sometimes called Calvary Clover because of the belief that Christ's blood splashed upon the leaves of the plant, which was said to have grown at the foot of the Cross. The spiky spirals look like small green cogs – hence its other name of Cogweed.

CONTROL Once this plant has seeded in your garden you will have it for several seasons, but it is easily pulled up by hand and like all the pea-flowers, is a good source of nitrogen in the compost-heap.

Spotted Medick

Yellow Melilot, King's Clover, Hay Flowers

Melilotis altissima Thuill

A Erect and branched
AH 60–120 cm (24–48 in)
L Leaflets oblong, parallel-sided
F Yellow. June to August
H Waste places and woods
Biennial

A tall plant, resembling a sparse-flowered, spindly lupin, which likes chalky or gravelly soils. It contains the sweet-smelling substance coumarin, found in many wild flowers.

The ancient Egyptians used Melilot as a remedy for ear-ache, and their magicians named it in spells to promote longevity. The sorcerers of the Middle Ages continued to hold it in high esteem, and added it to their potions and philtres. So many magical properties were eventually attributed to the herb that people began to be disappointed in its healing powers and gradually its use as a medicinal herb diminished. However, it is still considered to have great curative powers for those patient enough to use it.

USES When dried, it can be laid among furs to repel moths and impart a long-lasting, pleasing fragrance. It is also an excellent addition to a herb pillow.

For rheumatic pain, an infusion (which smells delightful) may be made by adding about 1½ oz (45 g) of dried flowering tips to 2 pt (1.14 l) of boiling water, leaving to infuse for quarter of an hour. Take a cupful last thing at night and another first thing in the morning and two more during the day. This infusion (used cold on a compress) can also relieve inflamed eyelids.

The plant may be made into a salve or ointment for external use on boils, carbuncles and other skin eruptions. A concentrate of the plant has been used as an anti-coagulant but this should not be undertaken without qualified homoeopathic advice, as large concentrated doses of the plant are dangerous.

CONTROL Easily weeded out by hand.

Yellow Melilot

PAPILIONACEAE Pea Family

Suckling Clover, Lesser Yellow Clover, Shamrock

Trifolium dubium Sibth

A Slender, procumbent or ascending
AH 4–25 cm (1½–10 in)
L Leaflets pinnately trifoliate
F Heads axillary, 10–26 flowered. May to October
L Grassy places
Annual

This wire-stemmed pest of lawns and drives forms a netted mat of stems which lie flat to the ground and remain cheerfully green even when the summer sun has burnt the rest of the lawn to coconut-matting. This small-leaved plant is the Shamrock of Ireland and St Patrick's Day (17 March).

CONTROL As the wiry stems are so strong, use a fine-tined rake to pull them out. Try to catch the plant before it flowers, as only a few seeds coming to maturity will mean repeating the whole performance next year.

PAPILIONACEAE Pea Family

Birdsfoot Trefoil, Bacon-and-eggs, Butter-and-eggs, Cuckoo's Stockings, Crow-toes, Lady's Slipper, Rosy Morn, God Almighty's Flower

Lotus corniculatus L.

A Decumbent
AH 10–40 cm (4–16 in)
L Leaflets obovate
F Yellow, streaked or tipped orange or red. June to September
H Pastures, roadsides
Perennial

This is the most familiar of the trefoils, and is sometimes found in the long grass of large old gardens. It has been given very many local names, some pretty, such as Rosy Morn and Love Entangled and some the reverse, such as Devil's Claws and Devil's Fingers. Plants noticed from earliest days usually have contrasting names. If it was called after the Devil – Cuckoo's Stockings – or one of his creatures, like Robin Good-fellow, Tom Thumb or Jack-Jump-About, then it was also named for God or the Virgin Mary – God Almighty's Flower and the universally-known, Lady's Slipper.

CONTROL Hand-weed if it should appear in a flower border or among the vegetables, but it does no harm among the long grass of a wild garden and can be happily left.

ROSACEAE Rose Family

Silverweed, Prince's Feathers, Goose Grey, Midsummer Silver, Silver Fern, Traveller's Ease, Crampweed

Potentilla anserina L.

A Creeping
AH 20 cm (8 in)
L Silvery, silky, pinnate
R Stoloniferous
F Yellow, 5 petalled, solitary. June to August
H Waste places, damp pastures, roadsides
Perennial

Silverweed has always been a plant of waysides and verges, and its exquisite leaf is now too often seen covered with roadside dust. In the past its form and shape were carved with loving care on the stalls and pew-ends of churches and cathedrals. Traveller's Ease was a name given to it because the foot-travellers of yester-year – pedlars, herdsmen, runners, packmen and soldiers – would stop to fill their boots with the fresh, cool leaves.

USES The plant contains tannin and up to the beginning of this century it was distilled into a lotion to heal and beautify the skin when sunburnt (a tan was considered a disgrace in those days) and to take away freckles, spots and pimples (for this last, the leaves were boiled up in wine-vinegar).

The leaves are beautiful in flower arrangements, and will grow very much bigger in good garden soil.

Potentilla means 'potent' and potent this beautiful little plant is. It may be used externally (2 oz (55 g) of the plant to 1 pt (0.6 l) water) as an infusion to arrest the bleeding of haemorrhoids and, used in weaker strength (1 oz (30 g) of the plant to 1 pt (0.6 l) water), may be taken internally at the same time. This infusion can be used as a gargle for sore throats (sweetened with honey) and a tablespoonful of the dried, powdered herb taken as a medicine every 4 hours.

On the Continent the herb has been used as a

Suckling Clover

Birdsfoot Trefoil

Silverweed

Creeping Cinquefoil

33

remedy for lockjaw, and the old herbals used Silverweed as a cure for ulcers in the mouth, as a remedy for the stone or gravel, and 'to heal wounds of the privy parts'.

CONTROL If there are no flower arrangers in your household to protect this plant, and you wish to be rid of it because of its invasive habits, the central parent plant should be found and rooted up, and then all the other offsets will come up too.

ROSACEAE Rose Family

Creeping Cinquefoil, Five-finger Grass
Potentilla reptans L.

A   Trailing, prostrate
AH 14 cm (6 in)
L   Palmate with 5 dentate leaflets
R   Stoloniferous, rooting at the nodes
F   Solitary, 5 petalled, yellow. June to September
H   Roadsides, waste places
    Perennial

Cinquefoil is an old herb, full of mystery and magic, which matches the charm of its name. It was credited with supernatural powers, and was an essential ingredient in love divination. According to Bacon, frogs liked to sit on this plant – 'the toad will be much under Sage, frogs will be in Cinquefoil'. The plant protects its fragile blooms in bad weather by contracting the leaves so that they curve over and shelter the flower.

USES The plant was one of the ingredients in an old fisherman's bait which was supposed to ensure a good catch – boil corn in Thyme and Marjoram water, and mix with Stinging Nettles, Cinquefoil and the juice of the Houseleek.

The plant is attributed with a long list of medicinal virtues. It is an astringent and a febrifuge. In former days the roots were used to cure marsh-agues and fevers, but in these more modern times they are used for a gargle and a mouthwash, and as a remedy for diarrhoea.

In America, the bark of the roots is used to stop nosebleeding. For this, the roots should be dug in April, and dried immediately (the outer bark only is used). Make a decoction as follows: take 1½ oz (45 g) of the dried root to 2 pt (1.14 l) of water, and boil it down to 1 pt (0.6 l). This may also be used to bathe tired eyes.

The crushed juice of the fresh roots, mixed with the crumbs of boiled wheat bread, is a styptic.

CONTROL The plant will make a net of runners and it is essential to find the parent plants and root these up, so that all the progeny will come up too. The leaves and whole plants are very good in the compost-heap.

ROSACEAE Rose Family

Wood Avens, Herb Bennett (Benet), Blessed Herb, Colewort, Star of the Earth
Geum urbanum L.

A   Erect and branching
AH 20–60 cm (8½–24 in)
L   Downy, pinnate
R   Rhizome
F   Yellow, erect, few. June to August
H   Woods, shady places
    Perennial

The name Herb Bennett is a corruption of the Latin *herba benedicta*, 'the blessed herb'. The roots have a pleasantly sweet and spicy smell like cloves and an old manuscript of 1491 says that if a root of Herb Bennett is in the house the Devil will flee from it, 'so it is blessed above all other herbs'.

This is a rather uninteresting-looking plant, and one feels that the petals of the flowers should be as large as those of some of its near relatives. The hooked burrs will transfer themselves eagerly to your clothes when out walking, and if you find Herb Bennett in your garden where none was before, you – or your dog – may have brought it back from a walk in the woods. This plant occurs as an ecclesiastical decoration from the end of the thirteenth century, and it is thought to symbolize the Holy Trinity and the five wounds of Christ in its trefoiled leaf and five, bright golden petals.

USES Dig the roots up in the autumn, dry them thoroughly and use them, as Gerard did in the sixteenth century, as a moth-repellent.

This plant has been used as a remedy since earliest times for colic, diarrhoea, dysentery, circulatory disorders, debility and liver malfunctions. It was taken as a tonic wine, which was made as follows: boil 1½–2 oz (45–55 g) of root in 2 pt (1.14 l) of red wine for 10 minutes. Allow to cool, then strain. Dose: 2 or 3 wineglassfuls per day.

This may be used as a digestive, a tonic and a depurative, and is deliciously fragrant. This tonic wine was considered a cure-all, and in the nineteenth cen-

tury was said to be 'good for the heart and clears the eyes, nose, teeth, brain and heart of anything that should not be there'.

Another old remedy of 1708 is said to strengthen the brain! – 'To the roots of Avens and Masterwort 1 fluid ounce of the leaves of Clary, Thyme and Marum of Cortufus, equal parts of Paeony – feeds 2½ drachms of Lavender Flowers and Lily of the valley, 1 drop of Tartar vitriolated 1 drachm; boil them in a sufficient quantity of fair water to ½ pint and make an Apozem for three Doses, adding to each Dose of the syrup of Baum 1 fluid ounce.'

CONTROL Herb Bennett has a habit of growing in among the roots of choicer plants, and is quite difficult to eradicate, especially if it is sharing a rock cranny with something less robust. Use a spot weedkiller in this instance or, if you are afraid of damage to the more delicate plant, keep cutting off all the top growth of the lustier Herb Bennett, until it eventually disappears.

ROSACEAE Rose Family

Agrimony, Cockleburr, Sticklewort, Sweethearts, Fairy's Wand
Agrimonia eupatoria L.

A   Erect
AH 30–60 cm (12–24 in)
L   Pinnate, with smaller pairs of leaflets between main ones, serrate, dark green
F   Numerous, yellow. June to August
H   Roadsides and hedges
    Perennial

Agrimony was one of the ingredients of 'Arquebusade water' which was compounded against wounds inflicted by the arquebus, or hand-gun. This remedy was spoken of by Philip de Comines in his account of the battle of Morat in 1476. In France, *Eau d'Arquebusade* is still carefully made from several aromatic herbs and externally applied to bruises and sprains.

Chaucer knew of the plant, which he called Egrimoyne – a corruption of 'Argemone', a name given by the ancient Greeks to herbs that healed the eyes. Gerard said that 'a decoction of the leaves is good for them that have naughty livers' and Pliny called it 'a herb of princely authoritie'.

This is a herb with magical associations – an old remedy for haemorrhages was a compound of Agrimony, frogs and human blood (a preparation discontinued in modern medicine!) – and, according to an eleventh-century verse, it was supposed to induce deep sleep:

> If it be leyd under mann's heed,
> He shal sleepyn as he were deed;
> He shal never drede ne wakyn
> Till from under his heed it be takyn.

The plant seeds itself by means of the small hooked burrs that cling with fierce determination to human garments and the fur of animals – if Agrimony springs up in your garden, you may well have brought it there yourself.

USES A useful decoction to soothe the throat may be made as follows: take about 4 oz (115 g) leaves to about 2 pt (1.14 l) water, and reduce by boiling to two-thirds of the original volume. Add ¼ oz (7 g) clear honey and a few drops of vinegar. Gargle with this liquid (strained) 4 or 5 times a day. Use also for stomatitis, gingivitis, sore throat and pharyngeal quinsy.

For slow-to-heal sores and ulcerated varicose veins and bruises, boil 8 oz (225 g) dried leaves in 2 pt (1.14 l) of red wine for about 5 minutes. Keep covered and leave to infuse until cool. Strain. Clean the wounds with this preparation, and then apply as clean compresses and bandage – renew as often as necessary. Other medicinal actions of the plant are tonic, astringent and diuretic.

CONTROL Not a common plant of the regularly tended vegetable garden, but quite likely to spring up at the edge of paths and drives. Root up by hand and use as compost.

Wood Avens

Agrimony

Biting Stonecrop

Large-flowered Evening Primrose

CRASSULACEAE Stonecrop Family

Biting Stonecrop, Wall Pepper, Golden Moss, Sedum

Sedum acre L.

A   Mat-like
AH 2–10 cm (1–4 in)
L   Glabrous, evergreen, imbricate, ovoid, trigonous, obtuse, sessile
F   Bright yellow, many. June to July
H   Walls, dunes, shingle
    Perennial

This plant often covers dry walls with blankets of blazing yellow, when the green mats of Biting Stonecrop are completely hidden by the hundreds of bright yellow, starry flowers which, alas, do not last more than a month.

This is a plant that was once planted on cottage roofs – particularly thatched ones – in order to ward off witches, lightning and evil spells, and Pliny claimed it as a cure for insomniacs if wrapped in a black cloth and surreptitiously placed under the pillow of the sleepless one.

The reddish colour that the stonecrops and other succulents develop in a hot summer is caused by the presence of anthcyanine, which is formed by the sap to protect the delicate inner tissues from the scorching sun.

Biting Stonecrop is well-named. Its acrid and burning juices can blister the skin and the plant should be handled with care.

CONTROL This is a hardy flower which only needs sun to survive. When it has grown into a 'mat', cut it away with shears. Like all stonecrops, dropped fragments will take root, so care should be taken to sweep up all the chopped-off pieces.

ONAGRACEAE Willowherb Family

Common Evening Primrose, Field Primrose, Tree Primrose, Fever Plant, Night Willowherb, Wild Beet

Oenothera biennis L.

A   Erect and robust
AH 50–100 cm (20–40 in)
L   Lanceolate to ovate-lanceolate, denticulate, green midrib, turning reddish
R   Fleshy

F   Yellow, smaller than *O. erythrosepala*. June to September
H   Waste places
    Biennial

This plant is native to North America and was first introduced to European gardens early in the seventeenth century. It is now a vigorous addition to the list of garden escapes. (It is not in any way related to the spring-flowering Primrose.)

If a garden is big enough, a large clump of these showy, clear-yellow flowers makes a handsome addition to fading, late-summer borders. The large delicately-scented flowers slowly unfold in the early evening; these flowers die the next day, to be replaced by others. Grandiflora or Lamarkiana are the best varieties to grow, as these have larger, finer flowers of the clearest yellow. Once one clump is allowed to seed, there will never be any lack of seedlings. Dig these up early in the first year, taking care not to damage the long root, and plant them in a mass.

USES The stalks of seed-heads, if cut and dried, make interesting winter dried-flower decoration.

An infusion for coughs may be made by adding about $\frac{1}{6}$ fl oz (5 ml) crushed plant to about $\frac{1}{2}$ pt (0.3 l) boiling water. The dose is about 2 fl oz (60 ml) 4 times a day.

The plant has also been used to alleviate depression and made into an ointment to cure skin rashes and irritations. Research is currently being done into the plant's anti-coagulant properties.

CONTROL All the adult plants should be destroyed before they seed. The small seeds can fall into a crack in a path or established rockery and a fine, tall plant will result, whose roots will be completely inaccessible. As it is normally a biennial, the shearing-off of the leafy parts may be enough to kill the plant, but sometimes it will try again in the third year, when spot weedkilling is the only remedy.

ONAGRACEAE Willowherb Family

Large-flowered Evening Primrose
Oenothera erythrosepala Borbás

A   Erect and robust
AH   50–100 cm (20–40 in)
L   Pubescent, lanceolate, crinkled with white or red midrib
F   Yellow, large, scented. June to September
H   Waste places
    Biennial

This variety of the evening primrose is on the increase whereas *Oenothera biennis*, for so long a common visitor to the garden, is on the decline, probably because it has been so ruthlessly destroyed. This variety has larger flowers and conspicuous red spots on the stem which *O. biennis* lacks.

CONTROL   As for the Common Evening Primrose.

UMBELLIFERAE Carrot Family

Alexanders, Black Lovage, Black Potherb
Smyrnium olusatrum L.

A   Stout and bushy
AH   50–150 cm (20–60 in)
L   Yellow-green, celery-like
F   Yellowish. April to June.
H   Cliffs, waste and stony places
    Perennial

This plant is indigenous to the Mediterranean shores and was introduced to northern Europe by the Romans and thereafter cultivated as a vegetable in monastery gardens. The crushed leaves and stems have a very strong and pungent scent. It is not very common except by the sea, and sometimes on chalk cliffs, but it is a great garden-invader, and on the bird-sanctuary island of Steep Holm in the Bristol Channel, it has smothered all other plant life except the Gorse and the Privet.

USES   The young stems can be eaten raw, like celery, or boiled, in which case boil them twice in salted water to mitigate the strong flavour.

CONTROL   It seeds itself everywhere so hand-weed before the seeds can form. The whole plant will come up easily if the ground is damp and will make a useful addition to the compost-heap.

UMBELLIFERAE Carrot Family

Fennel
Foeniculum vulgare Mill

A   Feathery
AH   60–130 cm (24–52 in)
L   Glaucous, much divided, aromatic
F   Yellow. July to September
H   Cliffs and walls
    Perennial

This very aromatic plant has been used medicinally and in the kitchen since earliest times. The Romans valued it, and it was a commercial crop in Spain as early as 961.

Fennel was one of the ingredients of the nauseous 'liquorice powder', a dreadful greenish compound formerly administered to unwilling children as a spring tonic. But, mixed into a syrup together with sodium bicarbonate, Fennel is still an important constituent of the present-day 'gripe water' which is given to babies. Culpeper was convinced that the whole plant (leaves, seeds and roots) could be 'much used in drink or broth, to make people more lean that are too fat'.

On a warm summer's day Fennel gives off a strong aroma like aniseed, which can be detected some distance from the plant, which then is instantly recognizable (whether in flower or not) by its mass of fine, green, hair-like foliage.

The plant used to be hung in the open rafters of old houses as a charm against evil, and it was hung over the door-lintel on Midsummer's-Eve in order to protect the inhabitants from witchcraft. Although usually regarded as a benign and protective herb, there is an old spell for summoning devils which requires Fennel, Coriander, Parsley, Hemlock, liquor of poppy, Sandal-wood and Henbane.

It is the food plant of one of the rarest and most beautiful butterflies, the Swallowtail (*Papilio machaon* L.).

Alexanders

Fennel

USES Powdered Fennel seeds can be used in kennels and catteries to repel fleas and lice.

It is an ideal herb to use with fish. Stuff some fresh-cut sprigs with butter, salt and black pepper into the belly of a mackerel and bake in foil for 25 minutes – it needs nothing more. Or salmon, poached in white wine, can be laid upon a bed of the bright green, aromatic sprays – infinitely better than limp lettuce.

Use it again with butter, salt and black pepper to cook conger-eel steaks (which *must* be fresh); bake them in foil to keep all the flavour in.

Fennel is good for conjunctivitis or swollen eyelids. Make an infusion by pouring 2 pt (1.14 l) water over 2 oz (55 g) seeds and allow to boil for 5 minutes in a saucepan. Place the pan on the table, drape a towel over the head and shoulders and sit over the steam with eyes closed. Use the same lotion, cold, to bathe forehead and temples to alleviate migraine or headaches.

An infusion to aid digestive difficulties and general sluggishness can be made as follows: to 2 pt (1.14 l) boiling water add 1 oz (30 g) dried seeds and leave to infuse for 10 minutes. Strain. Dose: 8 fl oz (225 ml) twice daily after a meal.

Fennel seeds are excellent for flatulence according to Sir John Harington in *The Englishman's Doctor* (1608):

> In Fennel-seed, this vertue you shall finde,
> Foorth of your lower parts to drive the winde.
> Of Fennel vertues foure they do recite,
> First, it hath power some poysons to expell,
> Next, burning Agues it will put to flight,
> The stomack it doth cleanse, and comfort well:
> And fourthly, it doth keepe and cleanse the sight,
> And thus the seeds and hearbe doth both excell.

For twentieth-century flatulence, however, take 1 tsp dried seeds, boil in a cup of milk for 5 minutes, strain, and drink as hot as possible.

A decoction of the root is a diuretic, and is made as follows: add 1 oz (30 g) grated root to 2 pt (1.14 l) water. Boil for a few minutes and allow to cool. Strain. The same decoction may be drunk before meals to stimulate a flagging appetite.

CONTROL If the plant has seeded, you may find Fennel growing anywhere in your garden. It is such a useful and beautiful plant, with its plumes of feathery green, that it is worth transplanting into the herbaceous border as a background plant (particularly the variety that has dark-bronze leaves which may be eaten in the same way as the wild Fennel). However, if you have forests of Fennel and enough is enough, dig the roots up and eat them as a vegetable.

UMBELLIFERAE Carrot Family

Wild Parsnip, Field Parsnip, Madnip, Bird's-nest, Hat's-eye, Tank, Siser
Pastinaca sativa L.

A   Erect and bushy
AH   30–150 cm (12–60 in)
L   Pinnate, segments ovate, lobed and serrate
R   Tap-rooted
F   Yellow. July to August
H   Roadsides, waste land and on chalky or rich soil
     Biennial

Cultivated parsnips are descended from the wild variety which the Romans ate with considerable appreciation and enjoyment. Parsnips were at that time cultivated along the banks of the Rhine and the Emperor Tiberius Caesar was so fond of them that they were sent to him annually.

Cultivated parsnips are very good fodder for pigs and cows, and come conveniently at a time when grass is in short supply. They will however, revert to the wild state quite quickly, whereas Wild Parsnips can never be tidily tamed nor induced to grow big and fat.

This is the only yellow-flowered umbellifer to be found growing along the verges – Fennel has yellow flowers but is more choosy in its environment, preferring clean, salt breezes to the dusty pollution of our roads.

For wine-making enthusiasts, the effort of getting the stringy little roots out of the unyielding grit at the edge of the road is really not worthwhile when the cultivated parsnip is not an expensive vegetable.

Gardeners should note that parsnips do not like manure, and will grow forked tails if there is any in the top spit.

CONTROL The Wild Parsnip, as may be expected, has a tap-root, which comes up quite easily from damp garden soil. If the plant seeds, you will have parsnips in the paving for a year or two, but once the leaves are recognized, the young plant is easily rooted up.

Wild Parsnip

Creeping Jenny, Moneywort, Cornish Moneywort, Herb Twopence, Serpentaria, Twopenny Grass, Creeping Charlie, Yellow Myrtle

Lysimachia nummularia L.

A Creeping
L Ovate, paired
R Stoloniferous
F Yellow, axillary, sub-campanulate. June to August
H Moist hedgebanks
    Perennial

Moneywort was so called because the paired leaves along the trailing stems looked like rows of pennies – *nummularia* is from the Latin *nummulus* – 'money'.

It was sometimes called Serpentaria because of the old legend: 'If serpents be hurt or wounded, they do heale themselves with this herbe, whereupon came the name of Serpentaria.' (Gerard)

This is a useful plant with attractive leaves and flowers for odd damp corners where ground cover is needed. For more sunny places the cultivated variety *L. nummularia aurea* has delightful, trailing sprays of paired, yellow leaves, which look particularly well against the grey stone of crazy-paving.

Moneywort was used as a 'wound herb' in medieval times, and both decoctions and ointment were made from the plant. Gerard knew it as a cure for coughs – 'The herb boyled in wine, with a little honey or mead, prevaileth much against the cough in children, called the chinne-cough, but it should be chine-cough, for it doth make as it were the very chine-bone to shake.'

CONTROL The plant reproduces itself vegetatively by sending out long stolons which produce more leaf-stalks, buds and small roots. In winter, the leaf-stems

43

Creeping Jenny

Mullein

between parent and offset wither, and the young plant's roots survive to grow next year. If parent and progeny are not wanted in the garden, thorough digging in the autumn will sift out the new small plants, whose aerial parts completely disappear in winter.

SCROPHULARIACEAE Figwort Family

Mullein, Great Mullein, Flannel-flower, Flannel Leaf, Hag's Taper, Hedge Taper, Aaron's Rod, Candlewicks, Blanketleaf
Verbascum thapsus L.

A Erect spike
AH 30–200 cm (12–80 in)
L Obovate-lanceolate, covered with thick white wool
F Yellow, in a dense flowering spike. June to August
H Sunny banks, waste places, dry soil
Biennial

This tall, furry-leaved plant may appear anywhere in your garden and is clearly recognizable, even when young, by its rosette of soft, silvery-white leaves.

Ulysses was given a Mullein stalk by the gods to defend him against the wiles of the beautiful and evil Circe, and the flower-spike has long been considered to have great protective qualities. According to an old herbal – 'If a man beareth one twig of this wort, he will not be terrified by any awe, nor will a wild beast hurt him, or any evil coming near.'

Roman ladies used to dye their hair yellow with Mullein flowers which had been steeped in lye, and a Roman herbalist, Alexander Trallianus, recommended using the ashes of the plant to restore grey hair to its original colour.

In medieval times the flowering spike of the plant was dipped in oil, grease or suet to make a candle or taper, and the soft leaves were scraped of their down, which was dried to make tinder and lamp-wicks.

The seeds are reputed to have narcotic properties, and fish poachers sometimes throw them into the water

Lady's Bedstraw

Ragwort

to stupefy the fish, although too many seeds will poison them.

Mullein plants make good border flowers and need no staking. Dig them up when young and plant them at least 18 in (45 cm) apart with a blue *Veronica spicata* in front to hide some of the tremendous stem length.

USES Mullein has astringent, emollient and demulcent properties and is still cultivated in Ireland for pulmonary afflictions. A simple infusion can be made from 1 oz (30 g) plant to 1 pt (0.6 l) water (strain twice through cloth to remove the tiny hairs). Dose: 1 wineglassful as needed. This infusion may also help in cases of diarrhoea, and is believed to strengthen the bowels.

American Indians smoked the dried leaves, which may be used as the main constituent of asthma cigarettes, or smoked in a pipe to relieve smoker's cough.

In Europe, an infusion of the flowers is used to cure coughs. The liquid must be well strained to remove the many fine hairs among the flowers, which would cause great irritation to the mouth and throat. This infusion may be sweetened with honey.

For ear-ache and other ear infections, Mullein oil is recommended (made by soaking the Mullein flowers in olive-oil in a corked bottle, which should be left in the sun or in a warm place for a few days). Warm the strained oil and put one or two drops in the ear several times a day.

CONTROL It is very easily pulled up, particularly when young, and composted.

RUBIACEAE Bedstraw Family

Lady's Bedstraw, Yellow Bedstraw
Galium verum L.

A   Slender and creeping
AH 15–100 cm (6–40 in)
L   Linear, mucronate, 8–12 in a whorl, rough
R   Stoloniferous
F   Yellow, in a compound panicle. July to August
H   Hedgebanks
    Perennial

This plant's name is probably derived from medieval times, when ladies of rank stuffed it with other sweet-smelling herbs into their mattresses. However, another legend has it that Jesus was born upon hay with *Galium verum* in it, which immediately burst into golden blossom and was thereafter called Our Lady's Bedstraw.

The plant contains the oil, coumarin, which is found in Sweet Woodruff and new-mown hay. It was used in the past to curdle milk for cheese-making, and in the sixteenth century the plant was used to make a footbath for travellers.

Lady's Bedstraw likes chalky soil and fresh air and its golden froth of tiny flowers can be a pleasing adornment to the garden.

It is the food plant of the Elephant Hawk moth, *Deilephila elpenor* L.

USES An infusion of the chopped herb is a diuretic, good for kidney and bladder complaints. Take 2 tbsp herb to 2 pt (1.14 l) cold water, bring to the boil and simmer for 5 minutes. Allow to cool, then strain. Dose: 3 cupfuls a day, taken between meals.

CONTROL If it is not possible to dig down to the roots, keep cutting off the aerial parts; this treatment will eventually destroy the plant.

COMPOSITAE Daisy Family

Ragwort, Ragweed, Fairy Horse
Senecio jacobaea L.

A   Erect and sturdy
AH 30–150 cm (12–60 in)
L   Lyrate, pinnatifid, sinuate-toothed, dark green
F   Golden-yellow, in a large flat-topped corymb. June to October
H   Waste land, waysides and pastures
    Biennial
    Poisonous

This plant of grassland and waysides flowers and seeds quite undisturbed by grazing animals, although the caterpillars of the Cinnabar moth (*Callimorpha jacobaeae* L.) with their striped 'football-jerseys' are nearly always found wherever Ragwort grows. The plant causes toxic cirrhosis of the liver in sheep, cows and horses, and its presence on grazing land is a sign of bad husbandry.

Ragwort is the legendary plant of witches who were believed to gather the strong coarse stems to make their broomsticks. An old Irish name for the plant is Fairy Horse, and it is said that the little people of Scotland used to use the plant to fly from island to island. Robert Burns wrote in one of his poems:

Let Warlocks grim and wither'd Hags
Tell how wi' you on ragweed nags
They skim the muirs an' dizzy crags
Wi' wicked speed.

Culpeper said in 1653 that the plant was 'singularly good to heal green wounds, and to cleanse and heal all old and filthy ulcers in the privities ...' and in 1812 another old herbal recommended that the basal leaves be used as 'pultices' and applied externally to areas of pain in the joints – this was said to have a 'surprising effect'.

CONTROL As Culpeper said, the root 'is made of many fibres, whereby it is firmly fastened into the ground, and abides many years'. Constant shearing of the aerial parts will eventually discourage the plant if it is impossible to get at the roots, and its mineral content will certainly benefit the compost-heap.

COMPOSITAE Daisy Family

Groundsel
Senecio vulgaris L.

A Erect and bushy
AH 8–15 cm (3–6 in)
L Glabrous, pinnatifid with irregularly toothed lobes
F Flower-heads in dense terminal clusters throughout the year
H Cultivated ground and waste places
  Annual

This plant's name is derived from the Anglo-Saxon *Groundswelge* or *Grundeswilige* which means ground-swallower and, left to itself, Groundsel will do just that. The plant used to be grown as a fodder-crop for small farm animals, such as goats, rabbits, pigs and poultry, and it is a delicacy, given in moderation, for cage birds.

USES Groundsel is a safe purgative with no subsequent ill-effects. Administer the plant in the form of a weak infusion ($\frac{1}{16}$ oz (2 g) to 4 fl oz (115 ml) water).

An old-fashioned remedy for chapped skin is to pour boiling water on to the fresh plant and apply this lotion to the rough skin.

A decoction of Groundsel (2 oz (55 g) fresh plant to 2 pt (1.14 l) water – dose: 5 wineglassfuls a day) is said to regularize the menstrual cycle in young or old, and soothe the pain which often accompanies irregular periods.

Groundsel

47

CONTROL It is a valuable source of iron so, if you have no rabbits or canaries, the best place for it is in the compost-heap. It is very easily weeded out by hand.

COMPOSITAE Daisy Family

Coltsfoot, Horsehoof, Coughwort, Son-before-father, Ginger Root, Clay Weed, Dove Dock
Tussilago farfara L.

A Scaly, leafless when in flower
AH Up to 30 cm (12 in)
L Appearing alone, roundish, 5–12 lobed, white-felted
R Stoloniferous
F Bright pale yellow. February to April
H Embankments, cliffs and waste places
Perennial

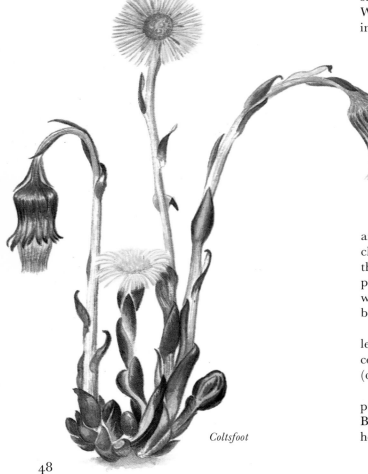

Coltsfoot

The yellow suns of Coltsfoot will appear long before the hoof-shaped leaves, and by the time that these appear in early summer, you may not associate them with the bright little flowers of a few months before. The stoloniferous stems can lie dormant for very long periods of time to spring into growth as soon as the ground is disturbed.

The soft down on the undersides of the leaves was used in the past by Scottish peasant-women to stuff their mattresses and birds often line their nests with the fluffy white clocks.

USES In former times, the plant was so much appreciated by apothecaries that the flower became the herbalists' shop sign.

The plant may be made into a herbal smoking mixture which has no harmful effect on the system and which relieves coughs, asthma, bronchitis and catarrh. Thoroughly dry the herbs and rub away all large pieces of stalk. Use Coltsfoot as a base, and to it add much smaller amounts of Thyme, Rosemary, Ground Ivy, Wood Betony and Chamomile flowers. Blend to suit individual taste.

The Romans knew Coltsfoot as a cure for asthma, and recommended burning the plant over a fire of charcoal; the smoke was then sucked into the mouth through a long, hollow reed and swallowed. The patient took a little wine between each inhalation, which must have made this particular treatment quite bearable, especially if continued for some time.

Another remedy for coughs is a decoction: 1 oz (30g) leaves to 2 pt (1.14 l) water; bring to the boil, and continue boiling until the liquid is reduced to 1 pt (0.6 l); strain and flavour. Dose: teacupfuls as needed.

Another recipe is to take 2 oz (55 g) dried plant to 2 pt (1.14 l) cold water. Allow to soak for a few minutes. Bring to the boil and then infuse for a quarter of an hour. Strain twice through a double thickness of cloth.

Corn Marigold

Flavour with lemon or blackcurrant juice and honey. Dose: 4 cupfuls a day. This is particularly good for a sore throat occasioned by continuous coughing. It has a restorative effect on the inflamed mucous membranes and may be taken with good results for laryngitis, colds, bronchitis and catarrh.

'Coltsfoot rock' may be bought in health shops in the form of long, yellow sticks; these are sucked slowly to alleviate coughs and tickling throats.

The leaves, freshly picked and crushed up with honey, may be used as a poultice for slow-to-heal external ulcers, wounds and erysipelas.

CONTROL Coltsfoot is very difficult to eradicate. It flourishes on heavy soil and can best be banished by adding straw, manure, sand, peat and composts of any kind to improve the soil. The roots burrow and run and must all be dug out, or the plant will continue to proliferate. Coltsfoot contains copper, calcium, iron, magnesium, potassium and sulphur and makes excellent compost, but it is best to burn the roots – they will probably not rot and may end up repopulating the entire garden.

COMPOSITAE Daisy Family

Corn Marigold, Gold
Chrysanthemum segetum L.

A Erect
AH 20–50 cm (8–20 in)
L Glabrous, glaucous, coarsely toothed
F Golden-yellow, large. June to August
H Acid arable soil, near roadworks
   Annual

This plant has long been hated by farmers. In a *Boke of Husbandrie* dated 1525, the plant was described as an 'yll wede' and even before that tenant farmers were ordered to uproot 'a certain plant called Gold' and courts were 'to amerce [fine or punish] careless tennants who do not weed it out before it comes to seed'.

However, in today's gardens, this plant can be a real asset. Its intense yellow flowers will light up a tired border until late summer and require no care.

CONTROL This plant is an annual, and so easily recognized when in flower that it can be removed to the compost-heap before it seeds. You will only be plagued with it if all the fields around are golden with its flowers, when it might understandably be called an 'yll wede'.

COMPOSITAE Daisy Family

Mugwort, Felon Herb, Sailor's Tobacco
Artemisia vulgaris L.

A Erect and bushy
AH 60–120 cm (24–48 in)
L Dark green above, white beneath, bipinnate
R Branching
F Dense racemose panicles, yellowish-brown. July to September
H Waste places
   Perennial

Mugwort, with its beautiful leaves, has flowers which, when they *are* noticed, resemble unopened buds or gone-over blossoms. It is a close relative of Wormwood (*A. absinthium*) and Tarragon (*A. dracunculus*).

This is an ancient plant, used as food in Stone Age times, and many legends have grown up around it. One old saying was: 'Eat Muggins in May and escape consumption, poison, fire, bills, beasts and other disorderly besetments.' Mugwort was generally regarded as a beneficent herb and its presence in the house was said to ward off lightning, and, if placed beside the door, prevent diseases from entering the house. It was also made into a cross and put on the roof to bring blessing to the house for the next year.

Travellers were told always to carry a sprig of Mugwort with them on their journeys to protect them from all harm. It was also one of the plants that were gathered – wet with dew – before sunrise on St John's Eve (23 June). The dew was considered to have especially magical powers on that day and at that time. In the evening fires were lit and all the herbs were hung in the smoke, which was considered to make them even more effective magically.

Artemesia (named after Artemis or Diana) is a plant especially for women and is supposed to help them in their menstrual cycles, the menopause, their confinements and even in conception. Its other, uglier name is possibly a corruption of 'midge-wort' because the aromatic leaves kept midges and mosquitoes away; alternatively, as it was used in beer-making (before the advent of hops), the name may be simply an association with beer-mugs.

Culpeper wrote in 1653 that the fresh juice of the herb was 'a special remedy upon the overmuch taking of opium' and Gerard said that 'Mugwort cureth the shakings of the joynts inclining to the Palsie'.

Mugwort

USES An infusion ($\frac{1}{4}$ oz (7 g) plant to 2 pt (1.14 l) water) has been used as a sedative and an anti-spasmodic, and recommended in cases of hysteria, nervous vomiting, chronic diarrhoea and epilepsy. Dose: 2 cups a day. It was in former times a remedy for gout, made by boiling a handful of leaves in a pint (0.6 l) of oil (such as olive-oil), until reduced to a third. A poultice was then made and applied to the painful area.

For irregular menstruation and lack of appetite a weaker infusion of 1–2 tbsp to 1 pt (0.6 l) water may be made. Use of the plant for these purposes should *not* be continued without the advice of a qualified homoeopathic practitioner.

The handsome leaves, with their silver undersides, are very beautiful and ideal for flower arrangements.

CONTROL Dig it up in winter or early spring.

COMPOSITAE Daisy Family

Nipplewort
Lapsana communis L.

A   Erect and thinly branching
AH 20–90 cm (8–36 in)
L   Lyrate-pinnatifid with a large terminal lobe, hairy
F   A corymbose panicle, pale yellow. July to September
H   Waysides, shady wood margins, waste places Annual

This plant was first called Nipplewort in 1588 by the German botanist Joachim Camerarius; he named it *Papillaris* 'because it is good to heale the ulcers of the nipples of women's breasts' and because the unopened flower-buds looked like nipples. Parkinson, one of the first great English botanists, translated *Papillaris* into 'Nipplewort'.

This plant survives because it is so inconspicuous and often goes entirely unnoticed until it flowers.

CONTROL It is very easily rooted up but it should be composted before it has time to seed or great patches of it will appear next year.

COMPOSITAE Daisy Family

Common Cat's Ear
Hypochoeris radicata L.

A   Erect flowering stems

Nipplewort

Common Cat's Ear

AH 20–60 cm (8–24 in)
L   Basal rosette, broadly oblong-lanceolate, hairy
R   Fleshy
F   Several from each leaf-rosette, flowers bright
    yellow. June to September
H   Meadows and waysides
    Perennial

This plant is very similar to several other yellow-flowered members of the *Compositae* family, but the Cat's Ear is perennial, with a neat rosette of basal leaves that hug the ground, and its flowering stems thicken slightly at the upper end (just below the flower-head).

Gerard said 'I find not anything set down at all either of (this plant's) nature or virtues, and therefore I forbeare to say anything of them, as a thing not necessarie to write of their faculties upon my owne conceit and imagination.'

USES The leaves are edible, and may be washed and used in salads, or cooked with other plants, as for spinach.

CONTROL This plant may star the grass verge outside your front gate and, as it has a long thin root, it may resist your efforts to get it out of the ground unless you use a tool such as a plantain-lifter.

COMPOSITAE Daisy Family

Bristly Ox-tongue
Picris echioides L.

A   Stiff and bristly
AH 30–90 cm (12–36 in)
L   Coarsely toothed and bristly ciliate, lanceolate,
    sinuate and covered with whitish pimples
F   Yellow. June to October
H   Roadsides and waste places
    Annual

At first glance, this pustule-covered plant looks diseased, but its pimply appearance is perfectly normal for this sturdy member of the great daisy family.

CONTROL Easily grasped and rooted out, once recognized by the whitish pimples on the leaves. Every downy seed means another plant next year, so compost it before the small white 'clocks' appear.

Bristly Ox-tongue

53

Goat's-beard

COMPOSITAE Daisy Family

Goat's-beard, Jack-go-to-bed-at-noon

Tragopogon pratensis L. Subsp. minor (Mill.) Wahlenb.

A Tall
AH 30–70 cm (12–28 in)
L Glaucous, linear-lanceolate
R Cylindrical tap-root
F Yellow. June to July
H Meadows, roadsides, waste places
  Annual or perennial

This is a tall, thin, inconspicuous plant, with flowers that never seem to be open. However, later in the year there will suddenly appear great, round, feathery 'clocks', which are more sturdily built than those of the dandelion, so that this 'clock' may be gently picked in order to admire the delicate strength of its construction. (You will have to blow quite hard to dislodge the seeds.)

The flower closes up at noon but, to compensate for this, it is often open at 2 am on a dry night. In former times field labourers were said to stop work for dinner when the Goat's-beard shut up for the day, but the flower is not as punctual as may be expected from its alternative name. It may even close up around 10 in the morning after fertilization has taken place.

USES If the seed-heads are sprayed with hair lacquer, they will stay together and can be used in flower arrangements.

The roots are edible, and can be cooked like carrots, adding butter before serving.

CONTROL When the great puff-balls of seeds appear, trap them in a paper bag and burn them.

COMPOSITAE Daisy Family

Common Sowthistle

Sonchus oleraceus L.

A Stout and erect
AH 20–150 cm (8–60 in)
L Ovate to runcinate-pinnatifid, glabrous, dull, never spinous
R Slender pale tap-root
F Yellow. June to August
H Cultivated soil, waysides, waste places
  Annual

Wherever man tills the soil, *Sonchus oleraceus* will pop up behind him. Neat though the herbaceous border may be in spring, by high summer the odd Sowthistle will have grown up, half-hidden in the centre of a clump of Phlox.

If you keep rabbits, Sowthistles make a particularly nourishing food for them and vets know the plant as a natural treatment for heart disorders, fevers and high blood pressure in animals. As may be imagined from the name, pigs eat this plant with especial relish.

Gerard said, 'The juice of these herbes doth coole and temper the heate of the fundament and privie parts' and Culpeper wrote enthusiastically of its virtues: 'The decoction of the leaves and stalks causes abundance of milk in nurses, and their children to be well-coloured' and 'Three spoonfuls of the juice taken, warmed in white wine, and some wine put thereto, causes women in travail to have so easy and speedy a delivery, that they may be able to walk presently after. It is wonderful good for women to wash their faces with, to clear the skin, and give it a lustre.'

In medieval times herbalists used Sowthistles instead of dandelions, as they believed the plants to have similar properties, and the early settlers in South Africa used the plant to make a poultice for healing external ulcers.

Sowthistle is not a thistle at all, though it is of the same natural order. Its seeds will drift a long way in the breeze so, although you may have completely cleared your garden of these sturdy visitors during the course of the year, they will cheerfully arrive again the year after – rather like relatives.

Sonchus oleraceus is the food plant of the Shark moth (*Cucullia umbratica* L.).

USES Sowthistles contain vitamin C and may be cooked and eaten like spinach, though it is best to add young nettles and dandelions, as the flavour of Sowthistle alone is a little too bland. Throw in a small amount of chopped chives and add a few drops of lemon juice, as well as the usual salt and pepper.

CONTROL Sowthistles are very easily rooted up by hand and are good in the compost-heap because of their high mineral content. It is as well to go round the garden specifically looking for these plants, as their succulent leaves provide lodging for garden pests, which may then be destroyed at the same time.

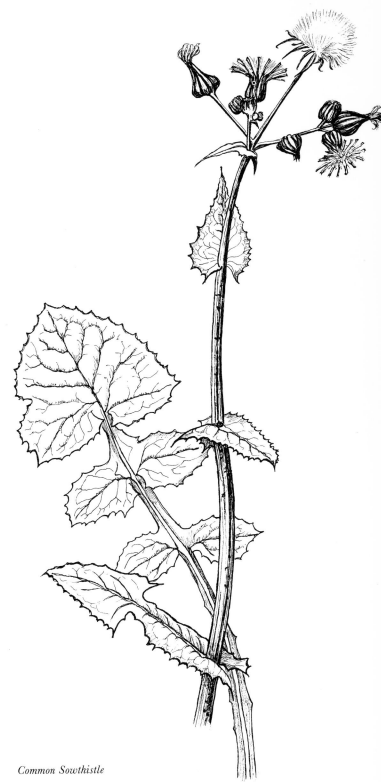

Common Sowthistle

Prickly Sowthistle

Prickly Sowthistle, Spiny Sowthistle
Sonchus asper (L.) Hill

A   Stout and branching
AH 20–150 cm (8–60 in)
L   Glossy dark green, crisped at the margin
F   Golden-yellow. June to August
H   Cultivated soil, waste places
    Annual

This plant is similar in height and appearance to the Common (smooth) Sowthistle, *S. oleraceus*, except for its leaves, which are soft-spined to the touch despite their slightly vicious appearance.

According to one of the old herbals, 'the milk that is taken from the stalks when they are broken, given in drink, is very beneficial to those that are short-winded and have a wheezing.' And another note, written in 1812 by a doctor, Sir John Hill, recommended – 'The leaves are to be used fresh gathered; a strong infusion of them works by urine, and opens obstructions; there are three or four other kinds of Sowthistle, and they all have the same virtues, but this has them most in perfection.' Obviously, the plant is not harmful, but it is not in general use today.

CONTROL The 'prickles' are deceptive and the plant is easily pulled out by hand. A valuable addition to the compost-heap.

Dandelion, Clock Flower, Piss-a-bed, Lion's-tooth, Blowball, Cankerwort, Milk Witch, Monk's Head, Irish Daisy, Priest's-crown
Taraxacum officinale Weber

A   Rosette-shaped
AH Up to 20 cm (8 in)
L   Usually pinnatifid, runcinate
R   Tap-root
F   Solitary, terminal, yellow. March to October
H   Meadows, lawns, waste places
    Perennial

Over one hundred different forms of dandelion have been recognized so if the ones that sneer at your lawnmower do not exactly match the description, it makes no difference – they are all equally tenacious of their environment and all equally prolific in their regenerative abilities.

There is no exactly defined reason for the name, which is derived from the French *dent de lion* (lion's tooth). Another name, Piss-a-bed or *Pissenlit* in French, is easily understood – over-indulgence in a salad of young dandelion leaves may indeed have that effect. It is a 'spring tonic' for the bowels and the French serve the leaves in salads in May, even in the heart of Paris.

Dandelion

Children tell the time by blowing at the perfect miracle of the seed-head, and their fingers are stained brown by the milky white juice from the stems. Girls sometimes use the 'clocks' to tell their marriage futures – the number of puffs needed to clear the seed-head of its fluffy seeds is the number of years that they will wait for a husband (some cheating is possible).

Grass verges ablaze with dandelion suns are a sign of warmer days to come, and there is a fable that the dandelion was created from the chariot-dust of Phoebus the Sun-God; as the sun rises the flower opens, and when it sets, the dandelion sleeps.

It has been said that if you rub yourself all over with the juice of the plant, you will be able to obtain everything that your heart desires, and you will have friends wherever you go.

Dandelion leaves are the food plant of the Clouded Buff moth, *Diacrisia sannio* L.

USES The young leaves may be used to make sandwiches, preferably with brown bread. (Salt or lemon-juice may be added if liked.) Tear the leaves into the size required instead of cutting them.

Dandelion wine is made as follows: fill a gallon (4.5 l) pot with the flower-heads. Pour over these 1 gal (4.5 l) boiling water. Stir thoroughly and cover with a thick clean cloth; leave for 3 days, stirring occasionally. Strain, boil for half an hour, and add 3½ lb (1.6 kg) sugar (loaf sugar is best), a small quantity of sliced preserved ginger, and the thinly sliced rings of a lemon and an orange, with all the pith removed. Put a little yeast on a piece of toast and add it to the brew, cover again and stand for 2 days to allow the yeast to 'work'. Strain again, pour into a cask and put in the bung – keep for about 2 months before bottling.

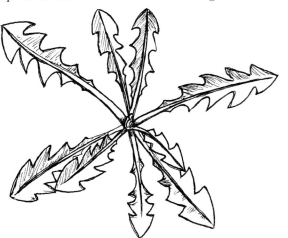

Dandelion roots may be used as a coffee substitute; dig up some large healthy plants from land that has never had an application of either pesticides or herbicides. Use only the fat single roots, not the thin forked ones. Wash them and allow to dry above the cooker, then spread them on a baking tray and roast them slowly until they are dark brown. Chop them into small pieces and grind as for coffee. This product, if carefully made, is excellent for the system, having none of the addictive and stimulating effects of real coffee.

The leaves are an excellent and very effective diuretic, having a slightly aperient effect on the system. A decoction taken internally stimulates bile secretion: 2 tbsp dried root to 1 pt (0.6 l) water, allow to macerate for 2 hours, then bring to the boil. Strain. Dose: this quantity during the day.

For sluggish livers, take 2 oz (55 g) freshly dug (and washed) sliced Dandelion root to 2 pt (1.14 l) water. Reduce to 1 pt (0.6 l) and add 1 oz (30 g) tincture of Horseradish. Dose: 3 fl oz (85 ml) per day.

An infusion of the leaves (easily found even in winter) is a tonic, stomachic, depurative, cholagogic, and a mild laxative in cases of chronic constipation. Take 2 tbsp plant to 1 cup boiling water, flavour as liked and drink this every day. Use the leaves during flowering time if possible, and the root in the autumn.

Dandelion tea is good for dyspepsia and is made as follows: take 1 oz (30 g) Dandelion plant to 1 pt (0.6 l) water, sweeten with honey. Drink several cups a day and one last thing at night, as this helps in cases of insomnia.

CONTROL The only way to a perfect lawn is to get down on your knees and lever out the whole of the root with a plantain-lifter, otherwise another plant will rise up to take the place of the snapped-off piece. Do this when the ground is wet.

COMPOSITAE Daisy Family

Marigold, Pot Marigold, Garden Marigold
Calendula officinalis L.

A   Straggling
AH Up to 45 cm (18 in)
L   Oblong-obovate and stickily hairy
F   Orange or yellow. May until frosts
H   A garden escape
    Annual, sometimes biennial

Marigold

The Marigold has been used in medicine since the twelfth century, and was once carried as a talisman, wrapped together with a wolf's tooth, in a bay leaf. An old herbal says that the flowers were so pleasing to look upon that they were thought to 'draw evil humours out of the head and strengthen the eyesight'. Today, this old-fashioned flower is regaining its popularity, and deservedly so, since it needs no cossetting and lasts well when cut. Allow the plants to run to seed and do not disturb the soil around that area. In the spring, hundreds of seedlings will appear, which will flower joyously from late May until October or even later. Dead-head regularly to prolong the flowering-time.

USES The flower petals, dried or fresh, may be used as a harmless colourant for rice and cakes instead of the more expensive saffron.

The Marigold is 'the homoeopathic antiseptic'. It has been used since earliest times as a remedy for the pains of gout and rheumatism; as a diuretic; as a remedy for jaundice; to help with painful menstruation

For external use, the plant is good for healing wounds without a scar, for soothing and healing burns, as an ointment in chiropody, and as a cure for warts. The plant is used fresh (chopped up as a simple poultice) or as a decoction for bathing the affected part (4 oz (115 g) to 2 pt (1.14 l) water) and the ointment is made by simmering 8 oz (225 g) flowers in 1 pt (0.6 l) water to which is added $\frac{3}{4}$ lb (340 g) pure lard. Boil gently until all water has gone; press, strain and store in a china, glass or earthenware jar with a natural cork or ground glass stopper.

CONTROL Easily rooted out and composted. If it is not wanted in the garden, do not allow even one flower to set seed.

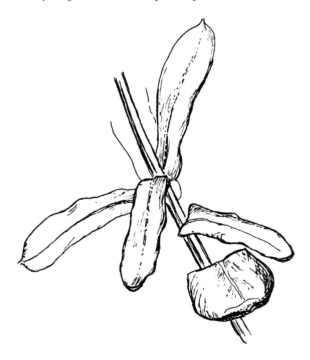

and menopausal problems. For all these ailments an infusion of fresh flowers should be made (2 oz (55 g) to 2 pt (1.14 l) water. Dose: 4 cupfuls a day). For the treatment of menstrual problems regular doses of this infusion should begin 1 week before the period is due.

White-flowered species

Field Penny Cress

CRUCIFERAE Cabbage Family

Field Penny Cress, Pennycress, Mithridite Mustard

Thlaspi arvense L.

A   Erect
AH 10–60 cm (4–24 in)
L   Glabrous, sinuate-toothed
R   Slender tap-root
F   4 petals, white. May to August
H   Roadsides, waste places, field-edges
    Annual

Instantly recognizable when in fruit by its flat, heart-shaped and very bright-green seed-pods, this plant was formerly used as an ingredient in an elaborate preparation called the 'Mithridite confection' which was considered an antidote to certain poisons. It thrives best in the poor soil of road verges and is thus not often found in gardens in sufficient quantity to be a nuisance.

CONTROL Easily weeded out by hand.

CRUCIFERAE Cabbage Family

Shepherd's Purse, Lady's Purse, Witches' Pouches, Rattle Pouches, Pick-purse, Mother's Heart, Case-weed, Clappede-pouch, St James' Weed, Toywort, Shepherd's Heart, Pepperplant, Pickpocket

Capsella bursa-pastoris (L.) Medic.

A   Rosette base with branching stems
AH 3–40 cm (1–16 in)
L   Variable, usually pinnatifid
F   4 petals, white. Flowering throughout the year
H   Cultivated land, waysides and waste places
    Annual

This little plant is extremely common. Its flowers are not very noticeable but it has distinctive shaped seed-pods from which it derives its name. The 'purse' (upside down) resembles the pouch which was worn at the waist by peasants to carry their food.

Shepherd's Purse was valued in the past for its medicinal properties as a blood-coagulant, and for use in chronic cases of menorrhagia.

USES The crushed herb (it should always be used fresh) may be made into an infusion and applied to bruises and strains, and to rheumatic joints: to 1 OZ

(30 g) fresh plant add 12 fl oz (340 ml) water, bring to the boil and reduce to 10 fl oz (285 ml).

CONTROL Hand-weed before the plant runs to seed. The new seedlings will appear in about 7 days as the ripe seeds need no period of repose.

Shepherd's Purse

CRUCIFERAE Cabbage Family

Hairy Bittercress, Small Bittercress
Cardamine hirsuta L.

A Rosette base with branching stems
AH 7–30 cm (3–12 in)
L Basal rosette, pinnate, dark green
R Slender tap-root
F 4 petals, white. March to September
H Found everywhere
   Annual

The neat and formal rosette of attractive leaves of the Hairy Bittercress may at first delude one into thinking it a self-sown seedling of some rightful garden occupant. However, its rate of growth and speed of reproduction make it a gardener's problem – it seems to populate a freshly-weeded flowerbed as soon as one's back is turned. In fact the plant is often introduced as a minute seedling, or even as a seed, via the container-pots from nurseries.

It is a tidy little plant, however, and pleasant to eat as a garnish to food or as a sandwich-filler.

CONTROL It is easily weeded out by hand but catch it before it seeds as it proliferates, by means of its explosive seed-pods, like Dragon's Teeth (*Tetragonolobulus Maritimus*).

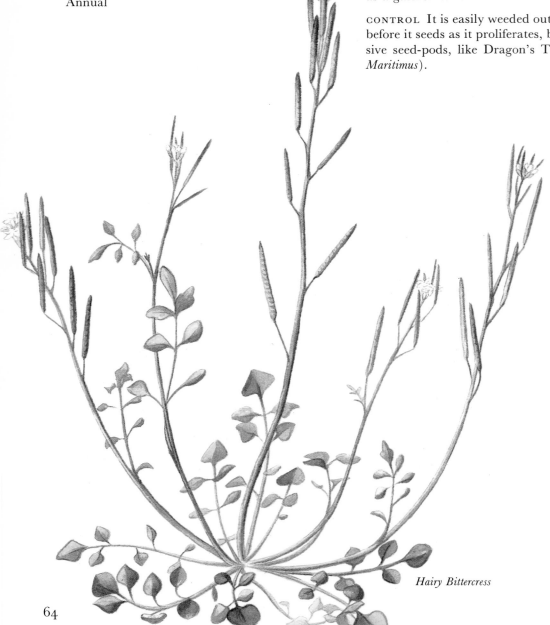

Hairy Bittercress

CRUCIFERAE Cabbage Family

Garlic Mustard, Poor Man's Mustard, Jack-by-the-hedge, Hedge Garlic, Sauce Alone

Alliaria petiolata (Bieb.) Cavara & Grande

A   Long-stalked, erect
AH 20–120 cm (8–48 in)
L   Reniform, cordate, pale green and smelling of garlic
R   Tap-root which smells of garlic
F   Petals white. April to June
H   Hedges, bases of walls, wood-edges, damp soil
Biennial

The bright, white flowers of this plant appear as if freshly washed and starched for spring, growing all along the hedgebanks in marshalled multitudes. The picked leaves smell of garlic, but not as strongly as the proper Wild Garlic (*Allium ursinum*), which gives off waves of goaty odour without ever being touched.

USES It makes a good sauce for lamb. In the early spring, chop the leaves with Hawthorn buds and a little mint, mix well with vinegar and sugar, and serve with the lamb as you would a mint sauce. It may be used alone (well-washed) in sandwiches, and tastes of mustard.

The plant can also be used as a spring diuretic for gout and rheumatism, and the expressed juice of the leaves taken alone, or mixed with honey and boiled to a syrup, is sometimes used to treat dropsy.

CONTROL Garlic Mustard only flourishes in damp soil, and, as it is tap-rooted, is easily pulled out of the ground.

CARYOPHYLLACEAE Pink Family

Common Mouse-ear

Cerastium holosteoides Fr. *(C. fontanum* L.)

A   Creeping
AH 15–45 cm (6–18 in)
L   Elliptical, dark greyish-green, covered with white hairs
F   White. April to September
H   Grassland, shingle, cultivated ground, waste places
Perennial

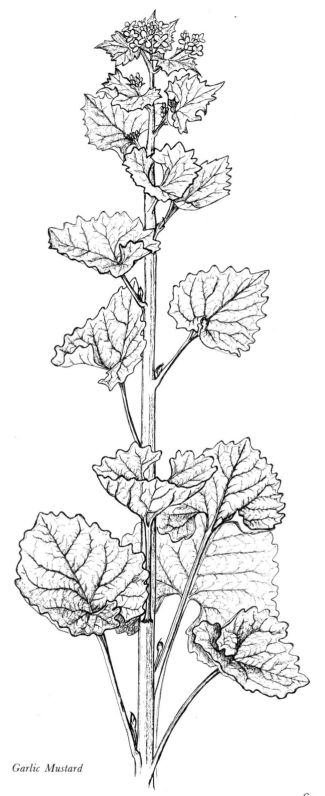

Garlic Mustard

Common Mouse-ear

One would never believe that this inconspicuous, furry dweller-in-lawns is related to the gorgeous Carnation and the Clove; its nearest garden relative is the Snow-in-Summer (*Cerastium tomentosum*) and the cloudy Gyp-sophila (*G. paniculata*).

CONTROL It is best to remove the plants with a plantain-lifter, unless the lawn is very large and the invaded areas are many and scattered, in which case use a chemical lawn weedkiller.

CARYOPHYLLACEAE Pink Family

Chickweed, Starweed, Starwort, Winterweed, Birdweed, Satinflower, Tongue Grass

Stellaria media L.

A   Diffuse
AH 5–40 cm (2–16 in)
L   Ovate-acute, glabrous, pale green
R   Slender tap-root
F   White. Throughout the year
H   Cultivated ground, meadows, waste places
    Annual

When people think of weeds one of the first names to spring to mind is Chickweed, because it is found everywhere and flowers all the year round if the winter is mild. It makes good bird food and, if there is enough of it, should be fed to young chicks – hence its name. The leaves are pale, clear green with a succulent appearance. Every night the opposite leaves lean towards each other, and the second-from-the-last pair closes up to protect the growing tip of the shoot.

It is the food plant of the Heart-and-Dart moth, *Agrotis exclamationis* L.

USES Chickweed tastes good and can be eaten raw. Wash and shake the leaves, add a squeeze of lemon, and use in sandwiches with tomatoes. Or cook it in the same manner as spinach, and, like spinach, gather twice as much as you think you will need. Add salt and pepper and some chopped Chives if liked.

For constipation, a tea can be made from the plant by pouring $\frac{1}{2}$ pt (0.3 l) boiling water over a handful of the washed leaves – add a little lemon for extra flavouring, and drink either hot or cold. This infusion (cooled and strained twice) is also very good to bathe tired eyes.

To make a poultice to relieve carbuncles, boils, and abscesses, place a handful of Chickweed in a small muslin bag, and boil for a minute or two. Apply as hot as may be borne to the affected part.

The plant may be made into a soothing bath to allay skin irritation from causes such as bee or insect stings, cuts, scratches and rashes: gather about 1 lb (455 g) Chickweed, and place in a preserving pan or similar vessel. Make an infusion by pouring about 7 pt (4 l) boiling water on to it. Allow to infuse for about half an hour, strain, and add to warm bath water. Do this every *other* day, and apply Chickweed ointment to the afflicted parts afterwards.

To make the ointment: gather about 2 handfuls of the fresh herb, wash and boil in about 4 oz (115 g) pure lard. Strain and allow to cool.

CONTROL It is no trouble to uproot this plant since its roots have only a tenuous hold in the soil.

Chickweed

Greater Stitchwort

CARYOPHYLLACEAE Pink Family

Greater Stitchwort, Star-flower, Adder's Meat, Thunder-flower
Stellaria holostea L.

A   Stems invisible among grass
AH 15–60 cm (6–24 in)
L   Lanceolate-acuminate
F   White, bifid. April to June
H   Hedgebanks, woodland
     Perennial

If you have inherited a neglected woodland garden, you may find bright scatterings of this flower in spring, though the white stars of Stitchwort are usually seen among the Bluebells in the hedgebanks and thickets. This plant is very weather-sensitive; to protect the pollen, the flowers droop forwards at the onset of rain.

Gerard said in 1597 that the flower got its name because it 'cured the paine in the side, stitches and such like' if it were drunk in wine mixed with the powder of acorns. One superstition made children afraid to pick the flowers 'in case adders bit them'.

CONTROL The plant is so brittle that it is easily pulled up by hand but do this when it is in flower and easily seen: if the seed-vessels have formed, the plant is almost invisible in long grass.

CARYOPHYLLACEAE Pink Family

Common Pearlwort, Beads
Sagina procumbens L.

A Tufted and with lateral shoots
AH Low
L Linear-subulate
F White, small. May to October
H Paths, lawns, gravel
   Perennial

This is so small and inconspicuous a plant that it may easily go unnoticed. It lives in lawns and neglected paths, and is so low-growing that it can flower and seed itself below the level of the mower-blades; these minute seeds are then carried to other parts of the garden on the soles of shoes.

It is a plant surrounded by many beliefs and superstitions, one of them being that it was the first plant that Christ stepped on when he rose from the dead. It was considered to bring luck and was placed over the door-lintel to keep fairies from coming into the house and stealing children for changelings. It was a charm for animals too – if a cow ate Pearlwort, she and her calf were deemed safe, and anyone who drank the cow's milk would be protected from harm. It was also a kissing charm for determined country maidens; if the girl kissed the man of her choice with a piece of Pearlwort in her mouth, he was bound to her for ever.

CONTROL If Pearlwort is all over your paths, try a chemical weedkiller specifically for paths. If the lawn, too, is thick with it, it usually means that the turf is neglected, even though closely mown. Pearlwort likes poor soil; improve the quality of the soil and the plant will vanish. A top dressing for the lawn which will help (two applications may be necessary) can be made with 3 parts sulphate of iron, 2 parts sulphate of ammonia, 1 part dried blood, and lime-free sand for distribution.

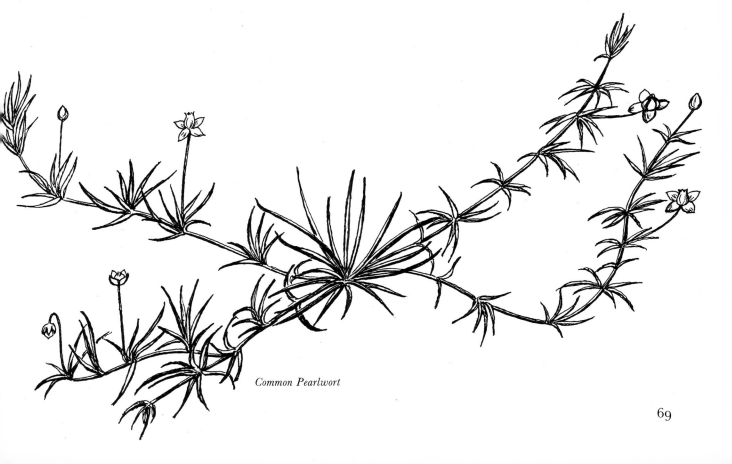

Common Pearlwort

69

PORTULACACEAE Purslane Family

Claytonia, Spring Beauty
Montia perfoliata Willd.

A   Erect
AH 10–30 cm (4–12 in)
L   2 united fleshy stem leaves
F   White. May to July
H   Disturbed ground, sandy soil
    Annual

Claytonia is a strange and interesting little plant, not obviously a weed. It originally came from the west coast of America over a hundred years ago, and it now grows very thickly in the Scilly Isles among the daffodils, though it is found elsewhere in Europe. The small bulbs, which are full of starch, were eaten as a vegetable by North American Indians.

CONTROL Very easily weeded out by hand if it has colonized a forgotten rose-bed.

Claytonia

PAPILIONACEAE Pea Family

White Clover, Dutch Clover, Kentish Clover, Honeystalk, White Trefoil, Purple Grass, Purplewort, Lamb Suckling, Shamrock, Trinity Leaf
Trifolium repens L.

A   Creeping and mat-forming
AH Up to 50 cm (20 in)
L   Leaflets with a whitish angled band
R   Rooting at stem nodes
F   White, numerous. June to September
H   Grassy places
    Perennial

This is the clover that the bees love best, and the phrase 'in clover' or 'living in clover' is a reminder of contented cows knee-deep in a clover field. The brownish flowers that hang down from the flower-head are those that have been fertilized by the hive bee on its honey-seeking visits. The honey is a clear amber colour, with a delicate flavour and fragrance.

In dry areas, clover leaves will always retain the typical, rich, emerald-green colouring, and if your grass regularly turns to coconut-matting in summer it might be worth considering a clover lawn. White Clover grows very quickly, rooting as it goes, and the nodules of nitrogen which form on the roots are an aid to the good health of the grasses in a mixed lawn.

Buttercups and clover do not get on well together, so if a green area is desired in the absence of grass, the buttercups will have to go.

CONTROL If a totally clover-and-weed-free lawn is desired, sulphate of ammonia may be used as herbicide. This has a corrosive action on all broad, flat and rough-leaved plants, and after it has been washed into the soil by rain, the nitrogenous salts that it contains activate the lawn-grasses into vigorous growth.

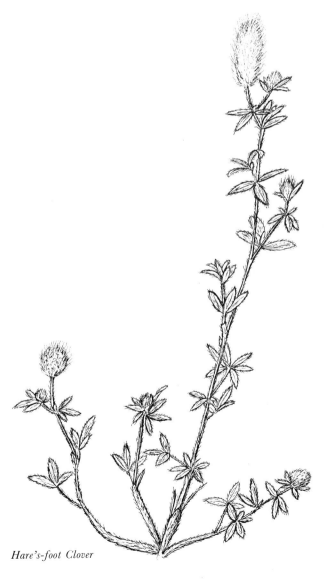

Hare's-foot Clover

PAPILIONACEAE Pea Family

Hare's-foot Clover, Rabbitfoot Clover
Trifolium arvense L.

A   Softly hairy
AH   10–20 cm (4–8 in)
L   Leaflets obovate, downy
F   Whitish-pink, cylindrical, downy. June to
     September
H   Sandy places
     Annual

This clover is distinguishable by the softness of its flower-heads, which have a silky furriness when in bloom. It appears only in sandy, gravelly areas.

CONTROL It is easily weeded out by hand and, like all clovers, makes a useful addition to the compost-heap. Always compost the whole plant unless the stems and roots are too woody.

ROSACEAE Rose Family

Wild Strawberry
Fragaria vesca L.

A   Trailing with arching, rooting runners
AH   5–30 cm (2–12 in)
L   Leaflets coarsely serrate-dentate, bright green
     above, pale beneath
F   White. April to July
H   Woods, rich soil
     Perennial

Izaak Walton said, 'Doubtless God could have made a better berry, but doubtless He never did'.

Wild Strawberries are an exquisite delicacy, particularly appreciated in France where it may be that the work of collecting these tiny fruits is always felt to be worth the effort to the discerning Gallic palate, which enjoys its *fraises du bois* every summer.

The first record of the Strawberry is found in a tenth-century plant list and the name has nothing to do with the modern practice of placing straw under the cultivated berries to keep them clean from mud-splashes. It derives in fact from the verb 'to strew', referring to the leafy runners which cover the ground.

A charming country custom was to thread the delicate berries on a grass-stem to make a bracelet.

USES For shining white teeth, clean them with the juice

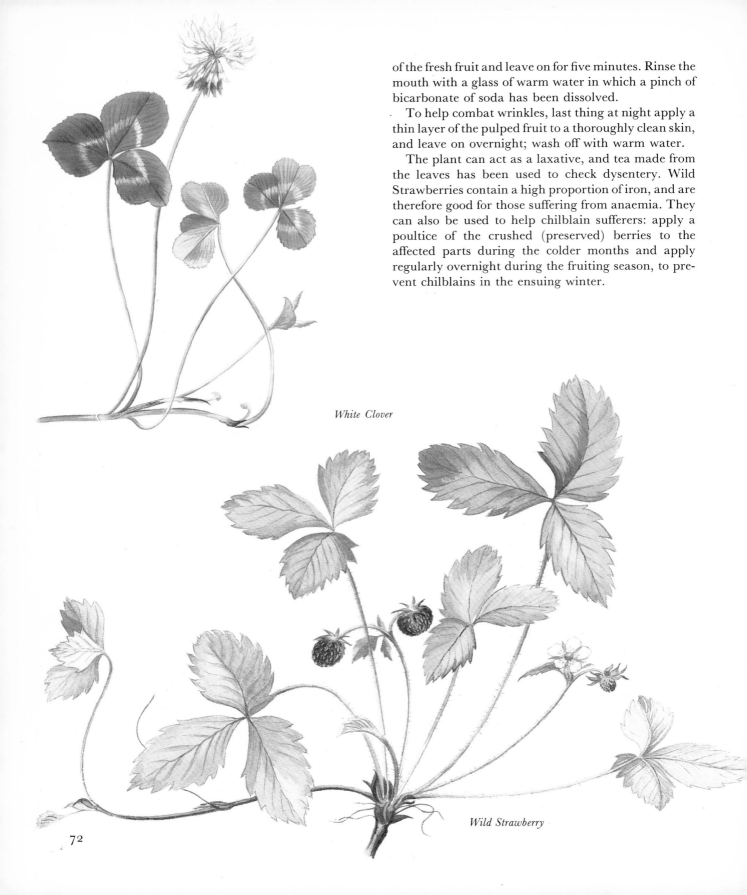

of the fresh fruit and leave on for five minutes. Rinse the mouth with a glass of warm water in which a pinch of bicarbonate of soda has been dissolved.

To help combat wrinkles, last thing at night apply a thin layer of the pulped fruit to a thoroughly clean skin, and leave on overnight; wash off with warm water.

The plant can act as a laxative, and tea made from the leaves has been used to check dysentery. Wild Strawberries contain a high proportion of iron, and are therefore good for those suffering from anaemia. They can also be used to help chilblain sufferers: apply a poultice of the crushed (preserved) berries to the affected parts during the colder months and apply regularly overnight during the fruiting season, to prevent chilblains in the ensuing winter.

White Clover

Wild Strawberry

CONTROL If you have Wild Strawberries in your garden, they are worth cultivating both for the visual pleasure they give and for their delicious taste. The plants like semi-shade and a rich soil with plenty of humus, and one plant, if it likes its habitat, will send out runners which in turn will do the same, and in three years there can be as many as two hundred plants.

CRASSULACEAE Stonecrop Family

White Stonecrop
Sedum album L.

A   Trailing and mat-forming
AH   7–15 cm (3–6 in)
L   Alternate, obovoid, terete, sessile, not spurred, glabrous
F   Many-flowered flat-topped cyme, white, pink-tinged. June to August
H   Walls and rocks
     Perennial

Culpeper used to call this plant the 'small Houseleek' and attributed to it all the virtues of the true Houseleek (*Sempervivum* family) which is reputed to protect the householder from lightning and evil spells. It is not a true wild plant, but it can invade the rock-garden or walls. The sedums are very difficult to kill, and have even been known to send out roots and new growth in a collector's drying press in a last desperate attempt at survival – such is their ability to withstand drought.

The plant was formerly made into a pickle, and the leaves were crushed and used as a cooling poultice for burns, scalds, insect-stings and even for painful haemorrhoids.

CONTROL Very easily trimmed back or removed when it exceeds its allotted boundary, but each small piece which is dropped will readily root again, so care should be taken to put all unwanted trimmings on the compost-heap.

White Stonecrop

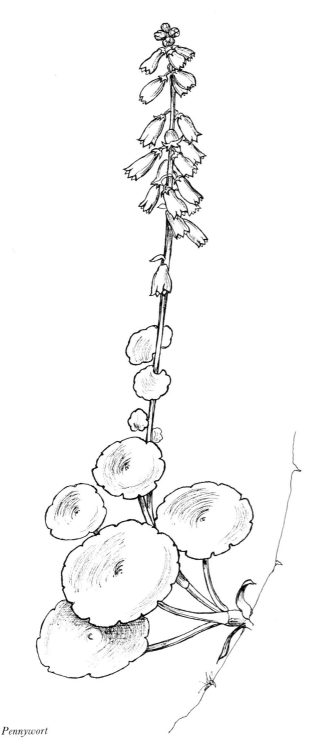

Pennywort

CRASSULACEAE Stonecrop Family

Pennywort, Navelwort, Kidneywort
Umbilicus rupestris (Salisb.) Dandy

A Mounds of flat, round shield-like leaves
AH 10–40 cm (4–16 in)
L Glabrous, succulent, orbicular, peltate, crenate
F Spires of greenish-white bells. June to August
H Damp walls and rocks
   Perennial

A familiar sight on damp walls, this neat plant with its umbrella-leaves once grew at Westminster Abbey 'over the doore that leadeth from Chaucers tombe to the old palace'. So relates Gerard in 1597, adding sadly that by his time it had disappeared. However, he also said that 'the juice of Wall Pennywort is good for kibed heels, being bathed therwith, and one or more of the leaves being laid vpon the heele', and records of the plant's medicinal uses are indeed met with in many old manuscripts. In the days when erysipelas was called St Anthony's Fire, the juice was used as a remedy and it was once used for epilepsy sufferers in the west of England, where the plant grows in great abundance. Today, it may be safely used as a cool poultice for burns, scalds and to soothe sore eyes.

CONTROL Wall Pennywort is no trouble to remove by hand if a tidy wall is required but, if it were difficult to grow or had a scarcity value, it would be seen in all the best rock-gardens since it is a most attractive plant with its spires of green-white bells rising from the thick, round, scalloped leaves.

UMBELLIFERAE Carrot Family

Cow Parsley, Queen Anne's Lace, Keck, Kex, Devil's Parsley, Wild Chervil
Anthriscus sylvestris (L.) Hoffm.

A Tall, feathery, branched
AH 60–100 cm (24–40 in)
L 2–3 pinnate, segmented, ovate and coarsely serrate
F Umbels of white flowers, 4–10 rays. April to June
H Hedgerows and waste places
   Biennial

Gerard said at the end of the sixteenth century that plants of this family once grew on 'the banks of St

Cow Parsley

James and Pickadilla', but now it is the most familiar of the 'umbrella-flowers' that whiten the grass verges in a foam of lace in early summer. It grows so quickly, so thickly, and so vigorously that the country lanes are made much narrower when it is in full bloom. The easiest way of differentiating this plant from others of the same species is to remember that this variety is earliest to flower, is the most common, and is very vigorous in its habit. It also has more flowers per plant than, in particular, its two most dangerous relatives – Hemlock (*Conium maculatum*) and Fool's Parsley (*Aethusa cynapium*). Hemlock can grow taller than a man, likes damp places and has a spotted stem. Fool's Parsley grows only to about 1 ft (30 cm) high and has a delicate spindly appearance, with whiskery bracts beneath the flowers.

This is an edible member of the great *Umbelliferae* family but care should be taken to identify it correctly. It is best picked in spring when identification is more easily ensured. It can be used in place of the cultivated Chervil in casseroles, soups and omelettes, or chopped and sprinkled on salads.

CONTROL Usually a biennial, the tall flowering plants may be scythed and left to dry. However, the whole area should be rotovated if it is to be cultivated, as the roots of Cow Parsley, though not individually tenacious, will be tangled with other established field dwellers, and all of them will resist the ordinary spade, however new and sharp.

UMBELLIFERAE Carrot Family

Ground Elder, Bishop's Weed, Goutweed, Herb Gerard, Garden Plague
Aegopodium podagraria L.

A   Erect and vigorous
AH 40–100 cm (15–40 in)
L   Glabrous, deltoid, irregularly serrate, bluntly triquetrous
R   Rhizomatous, far-creeping, white when young
F   Umbels white, rays 15–20. May to July
H   Waste places and gardens
    Perennial

Possibly the most pernicious weed in the garden, the tremendously long roots of Ground Elder creep under paths and start into lusty growth some distance away. Gerard's garden was probably just as plagued with it

Ground Elder

since he wrote, 'Herbe Gerard groweth of itself in gardens without setting or sowing and is so fruitful in its increase that when it hath once taken roote, it will hardly be gotten out againe, spoiling and getting every yeare more ground, to the annoying of better herbe.' The plant was brought to Britain from Europe by the monks of the Middle Ages as a pot-herb, a vegetable, and a medicinal plant and it was called 'Herb Gerard' after Saint Gerard to whom the brothers appealed for relief from the pains of gout.

USES A tea may be made by putting a handful of the bright green leaves into a clean teapot and pouring ½ pt (0.3 l) boiling water over them. Leave to infuse and keep covered. Single wineglassfuls several times a day will help to ease aching joints from gout or sciatica. These may be further eased by applying externally a

UMBELLIFERAE Carrot Family

Fool's Parsley
Aethusa cynapium L.

A   Branched and leafy
AH 5–120 cm (2–48 in)
L   Deltoid, segments ovate, pinnatifid, dark green
F   Umbels white, rays 10–20, 3–4 bracteoles on outer side of umbels. July to August
H   Cultivated ground
    Annual
    Poisonous

This umbellifer is like many of the others in height and

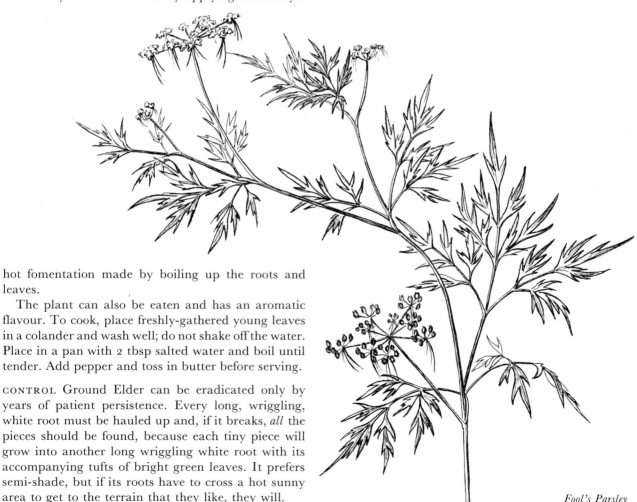

hot fomentation made by boiling up the roots and leaves.

The plant can also be eaten and has an aromatic flavour. To cook, place freshly-gathered young leaves in a colander and wash well; do not shake off the water. Place in a pan with 2 tbsp salted water and boil until tender. Add pepper and toss in butter before serving.

CONTROL Ground Elder can be eradicated only by years of patient persistence. Every long, wriggling, white root must be hauled up and, if it breaks, *all* the pieces should be found, because each tiny piece will grow into another long wriggling white root with its accompanying tufts of bright green leaves. It prefers semi-shade, but if its roots have to cross a hot sunny area to get to the terrain that they like, they will.

Fool's Parsley

Hogweed

78

habit, but is distinguished by the bracteoles (usually three) curving downwards from under the flower. It does not look much like Parsley when young, but since it is poisonous, it is as well to take extra care if the plant is common in your area or if you do not weed the herb bed very often. If in doubt, crush it and some true Parsley in separate hands and sniff. According to Gerard, Fool's Parsley has 'a naughtie smell' but at the time of flowering and seeding it just smells *green*, whereas Parsley has its own distinctive smell.

CONTROL Pull it up by hand and compost it, well before it seeds.

UMBELLIFERAE Carrot Family

Hogweed, Kecks, Cow Parsnip, Gipsy's Lace
Heracleum sphondylium L.

A   Stout, erect and bushy
AH 50–200 cm (2–80 in)
L   Variously lobed and pinnatisect, hispid
F   Umbels white, petals of outer flowers longer. June to September
H   Roadsides and hedges
    Biennial

This umbellifer is seen in hedgerows throughout the summer and is an altogether more substantial plant than the early-flowering umbellifer, Cow Parsley (*A. sylvestris*). A group of these plants looks very handsome in the wild garden, and the seed-heads look interesting throughout the winter. These also make attractive, tall flower arrangements. The plant is an excellent pig food, from which it derives its name.

USES The young shoots may be cut, tied into short bundles, and cooked as for asparagus. The coarse, hairy stems collect dust so pick plants away from the road verge.

CONTROL A large plant to remove in its adult state so try to identify it by its folded green leaves before it is fully grown.

UMBELLIFERAE Carrot Family

Giant Hogweed, Cartwheel Flower
Heracleum mantegazzianum Somm. & Lev.

A   Enormously tall
AH Up to 3.5 m (12 ft)
L   Leaves up to 1 m (3 ft), pinnately divided, shining green
R   Fanged
F   Umbels of white flowers, rays numerous. June to July
H   Streamsides and waste places
    Perennial

This is a spectacular giant of a plant that is impossible to ignore. Never handle it with bare hands and always wear a long-sleeved shirt or jacket as well as gloves to touch it because, on hot sunny days, the hairs on the stems exude a virulent substance that will cause blisters or a painful rash, especially on the more delicate skin of the inside forearm. (Children have been scarred around the mouth through using the hollow stems as pea-shooters.)

It is, however, a statuesque plant for the water's edge or the wild garden, if a corner can be found where the hairy leaves will not be brushed against. The huge cartwheels of the dried umbels can also be an impressive part of a flower arrangement.

CONTROL If it is wanted in the garden, the plant can be moved only in the winter when it is dormant, and its site should be marked earlier, as the leaves die down completely. It has a tremendous root, even in the second year, which must be dug out entirely and burned; any particle of root left will probably grow again, particularly in a wet summer.

POLYGONACEAE Dock Family

Japanese Knotweed, Mexican Bamboo
Polygonum cuspidatum Sieb & Zucc. *(Reynoutria japonica)*

A   Bushy, tall and vigorous
AH Up to 2.5 m (8 ft)
L   Ovate, cuspidate, truncate at base, deep green
R   Rhizomatous
F   Lax panicle of creamy flowers. August to October
H   Waste places
    Perennial

This tall, handsome, red-stemmed plant is now only too familiar. Formerly cultivated as an ornamental background plant introduced from Japan over a hundred years ago, it has spread rapidly. Once the first seedling has grown to maturity and established a good root formation, dozens more of the red-tinted offsets spring up through cracks in concrete or asphalt, needing no care or attention whatsoever, and forming a thick plantation in a very short time.

It is however a handsome plant, if it can be kept within bounds – the growing shoots will crack thin or old concrete, which must be reinforced if the plant is to be contained. If you are a brave gardener, and need to camouflage an ugly corner, this rampant plant with its bright green leaves, red stems and feathery white flowers will do the job more quickly than anything else. (It is related to that quick cover-up climber, *Polygonum baldschuanicum* or Russian Vine.)

CONTROL If the roots are under paving and inaccessible, constant cutting of the growing shoots is the only method of control. This must be done at least once a week in the spring, and more often in wet weather. If the garden is buried in a forest of Japanese Knotweed, chemical warfare is the only quick and (more or less) permanent answer.

Privet

OLEACEAE Olive Family

Privet
Ligustrum vulgare L.

A   Thin shrub
AH   Up to 5 m (16 ft)
L   Neat, lanceolate, dull green
F   Cone of cream-white flowers. June to August
H   Scrubland
    Poisonous

Privet is included here because of its greedy nature and the insidious way that the young plants grow to form impenetrable thickets. A Privet hedge will creep forward into and over flowerbeds, draining the nourishment from the soil around the hedge. It will also produce drifts of unobtrusive offspring, which are near-invisible under the shade of trees until the scented, white flower-cones appear, or the jet-black (poisonous) berries in late autumn.

The flowers of the Privet have a strong, unpleasant scent, and honey made from them has a fishy taint.

It is the food of the giant caterpillars of the Privet Hawk moth, *Sphinx ligustri* L.

CONTROL A neglected Privet hedge is full of fresh growth in the upper part but has bare lower stems. Try clipping the hedge hard until it tapers towards the top so that, seen end-on, the hedge will be a wedge or triangle, with the base at the bottom. This shaping will allow light, air and rain to get to the lower branches, which then may send out some branchlets to fill some of the gaps.

If an old-established hedge has to be removed, muscle power is the only answer. The soil in which the hedge grew will need replacing or, at the very least, comprehensive fertilizing.

CONVOLVULACEAE Bindweed Family

Hedge Bindweed, Wild Morning Glory, Bearbind, Devil's Guts, Old Man's Night Cap, Hedge-bell, With-wind, Withybind
Calystegia sepium (L.) R. Br

A   Twining stems in pairs
L   Ovate-cordate, sometimes sagittate
R   Long and penetrating
F   White, large, funnel-shaped. July to September
H   Hedges, gardens, waste places
    Perennial

Hedge Bindweed

This large wild flower is often considered the most vigorous weed in the garden and the most difficult to eradicate. In a few weeks of warm spring weather, its leafy top growth can completely strangle tender garden plants and its extensive root system spreads with horrifying rapidity in all types of soil.

An old witch's spell only to be undertaken during the three days before the new moon uses the plant to 'bind' its victim – an image, in mud or dough, of the person that is to be bewitched is baptized and bound with a length of the weed nine times (nine revolutions anti-clockwise) while chanting nine times 'I bind (Name) against/to (whatever action is not wanted/wanted). So shall it be'. The bound doll is then buried where the victim will walk.

USES The plant has traditional medicinal qualities. A gentle laxative can be made by infusing a handful of leaves (fresh or dried) in a cupful of boiling water for 5 minutes; strain. Drink about a quarter of an hour before breakfast.

A poultice of crushed, fresh leaves can be applied to a boil, which should break the next day.

CONTROL Hand-weeding is the only really effective method, but very deep and thorough digging is needed to make sure that all broken-off root portions are collected for subsequent burning. It is best to do the weeding on a damp day, when the roots are less likely to break. Every piece that is left will regenerate within weeks into another fine specimen of the genus. Where hand-weeding is impractical, constant cutting of the aerial growth is the only answer. (One way of killing the plant without digging it up is to unwind the stem from the host plant and rewind it in the opposite direction. It will slowly die from this treatment, though the sight of slowly yellowing leaves and the tangle of stiffening stems is an affront to a tidy gardener.)

SOLANACEAE Nightshade Family

Black Nightshade, Garden Nightshade, Poisonberry
Solanum nigrum L.

A   Bushy
AH Up to 60 cm (24 in)
L   Ovate or rhomboid, cuneate at base
F   Cymes of white flowers. July to September
H   Waste places, gardens
     Annual
     Poisonous

This inconspicuous, dull-green plant is named after its black berries which do not appear until late summer. As Culpeper wisely said, 'Do not mistake the Deadly Nightshade for this, if you know it not, you may then let them both alone.' Black Nightshade looks rather like a small, unformed potato-plant, to which it is closely related.

The berries are more poisonous than those of the Bittersweet, another member of the nightshade family, and care must be taken not to include the unripe green berries (which are the same size) when gathering peas. This may seem unlikely, but it has happened with

Black Nightshade

Thorn-apple

84

White Dead-nettle

enthusiastic volunteer help. The poison in the berries varies according to district, season and growth conditions, and what may be relatively harmless in one area may be fatal in another.

In remote parts of Europe, the leaves are placed in the cots of wakeful babies to promote sound sleep, and the plant is used medicinally, but only by qualified homoeopaths, as it contains solanine, which is toxic.

CONTROL This plant should be rooted out as soon as recognized. Always wear gloves to pull it up, and burn it afterwards.

SOLANACEAE Nightshade Family

Thorn-apple, Mad-apple, Jimson Weed, Jamestown Weed, Angel's Trumpets, Stinkwort, Stinkweed
Datura stramonium L.

A   Stout, erect, branched and evil-smelling
AH Up to 1 m (3 ft)
L   Ovate, sinuate-dentate, coarsely toothed, acuminate
F   White, tubular, acuminate lobes. July to October
H   Waste places
    Annual
    Poisonous

This coarse, shaggy-looking, malodorous plant seems to be happiest growing on a dry, hot railway embankment or forcing its way up through rusting remains in scrapyards. Its leaves are beloved by thrips and by late summer are often just a skeletal structure. The flowers do not open properly until evening (they are generally fertilized by moths) and they then have a sweet, tropical fragrance, which is strong enough to cause dizziness if inhaled deeply. However, the leaves of the Thorn-apple smell rank and fetid, even when undisturbed.

The seeds are exceedingly poisonous and their toxic properties are not destroyed by boiling or drying. Unfortunately, the half-ripe seeds have a sweetish flavour, and particular care should be taken with children, who may not be put off by the repellent and sinister appearance of the plant.

In the Middle Ages, and even up to the present day, Thorn-apple was one of the herbs used as incense during Black Masses. It has a powerful hallucinatory effect when burnt, particularly if Myrtle, Henbane, Rue and Deadly Nightshade are added.

Thorn-apple, when dried and chopped up, is used for making asthma-powders and cigarettes, which have a paralysing effect on the endings of the pulmonary branches, thus mitigating bronchial spasms. The plant is also made into ointments and plasters to alleviate pain from rheumatism, and reduce the swelling of abscesses.

CONTROL It is easily pulled out of the ground but always use gloves. (Should the juice of the plant get in your eyes, the pupils will become widely dilated.) The plant is better burnt than composted, since the ripe seeds can lie dormant for several years.

LABIATAE Labiate Family

White Dead-nettle, Adam-and-Eve-in-a-Bower
Lamium album L.

A   Nettle-shaped but soft and stingless
AH 20–60 cm (8–24 in)
L   Ovate, acuminate, coarsely serrate, hairy
R   Creeping rhizome
F   Tubular, greenish-white. May to December
H   Hedges, roadsides, waste places
    Perennial

This harmless plant is distinguished by the softness of its leaves and its green-white flowers, which are out for 7 months of the year. The White Dead-nettle nearly always grows in close proximity to Stinging Nettles, though it is not in any way related. The charming country name 'Adam-and-Eve-in-a-Bower' is explained by turning the flowering-stem upside down, whereupon the stamens will be seen nestling side by side, roofed by the bower of the green-white corolla.

USES A decoction of 2 tbsp plant to 2 pt (1.14 l) water is a useful remedy for diarrhoea, and may also be used to alleviate constipation. The lotion, strained, makes a cooling compress for burns.

CONTROL Hand-weed. White Dead-nettle has long running roots and may take some time to eradicate. The leaves are of great value in the compost-heap.

PLANTAGINACEAE Plantain Family

Greater Plantain, Common Plantain, Rat-tail Plantain, Bird's Meat, White Man's Foot

Plantago major L.

A Rosette-shaped
AH Up to 30 cm (12 in)
L Ovate or elliptic
R Tap-rooted
F Yellowish-white. May to August

Greater Plantain

H   Roadsides, cultivated ground, lawns
Perennial

Hated by gardeners who aim for perfect lawns, this plantain and its relatives have adapted themselves to the passage of the lawn-mower – flatter plants can seldom be found.

There is a sad legend of the Knight Hildebrand who went off to the Holy Wars, leaving his wife, the Lady Ermyntrude, to guard his castle. He asked her to watch beside the castle gate for his return and, patiently she waited, day after day, even after the tidings came that he had been slain in battle. In the spring the Lady Ermyntrude died from the effect of her long winter vigil beside the road and, soon afterwards, a little green plant that had never been seen before began to grow all along the road that led to the castle. The peasants said that the plantain was she, standing beside the road to make sure she saw her lord returning and be first to greet him.

Red Indians called this plant White Man's Foot because it was found wherever the early settlers had passed. The seeds are a nutritious treat for cage-birds, and it has been called Bird Plantain because of the fanciful story that once in seven years the plant becomes a bird to begin the search for cuckoos on the wing in order that it may serve them. The sight of wild birds rising up from the plants after feeding was probably the origin of this particular fable.

USES Valued throughout the ages as a herb of healing, plantain is famous as a 'wound herb' washed, crushed and applied as a poultice to cuts and grazes. It is also a remedy for insect stings and bites, sores, and varicose ulcers. To ease coughs, a decoction (4 oz (115 g) chopped leaves to 2 pt (1.14 l) water) can be made as follows: soak the leaves for 15 minutes in cold water, bring to the boil, boil for a few minutes and then allow to infuse for a further 15 minutes. Dose: 4 cupfuls a day; make fresh each day. This remedy is good for all diseases of the respiratory tract, whooping cough, bronchitis, and may also be used in cases of dysentery and diarrhoea. Make a decoction in the same way but use less of the plant (2 oz (55 g) to 2 pt (1.14 l) water) to bathe tired eyes and for cases of conjunctivitis.

For constipation, 2 tsp dried seeds soaked in a wine-glass of water for a few hours; drink the strained liquid before going to bed.

CONTROL No amount of normal tread seems to wear Plantains away and in the lawn the 'plate' of flat leaves will grow bigger and kill the grass surrounding it. The old-fashioned method of digging each plant out individually with a plantain-lifter is still the best, though 'spot' weedkilling may sometimes be necessary on paths and crazy-paving.

PLANTAGINACEAE Plantain Family

Ribwort Plantain, Kemps, Ribgrass, Buckhorn, Ripple Grass, Blackjacks, English Plantain
Plantago lanceolata L.

A   Erect spear-shaped leaves
AH Up to 15 cm (6 in)
L   Lanceolate, 3–5 nerved
R   Tap-rooted
F   Dull brownish flowers. April to August
H   Grassy places
Perennial

Country children play a game called 'Kemps' with the flower-heads of the plantain, in which the stems are struck against each other until one head falls off, signifying the beheading of a warrior. (The Old English word for soldier is *Cempa*.)

Culpeper speaks highly of the plantain as a general healing herb – 'The juice mixed with oil of roses and the temples and forehead anointed therewith, eases the pains of the head proceeding from heat and helps lunatic and frantic persons very much.' This plantain may be used in a similar manner to *Plantago major* for the same ailments.

CONTROL See under *Plantago major*, bearing in mind that this plantain's leaves are longer and thinner and do not form the flat rosette that hugs the ground with such determination.

Ribwort Plantain

89

Goose Grass

RUBIACEAE Bedstraw Family

Goose Grass, Common Cleavers, Clivers, Sticky Willie (Billy), Hayriff (Hairiff), Stick-a-back, Everlasting Friendship
Galium aparine L.

A Scrambling and diffusely branched
AH 15–120 cm (6–48 in)
L 6–8 in a whorl, linear, hooked, oblanceolate, mucronate
F Whitish in axillary cymes. June to August
H Hedges, waste places, shingle
  Annual

This plant has been fed to geese and newly-hatched goslings since the fifteenth century – hence its name. Another country use for the plant was to make a mat from the long stems, which acted as a sieve to strain cow's hairs out of new milk. The strange names Hairiff, Herif, Hayriffe or Eriffe are all derivatives of the old English word for a robber, because the plant 'steals' wool, hairs or feathers from anything that it touches.

Goose Grass belongs to a large and important plant family, among the members of which are *Coffae arabica* (from which coffee is obtained) and *Cinchona officinalis* from which the drug quinine is extracted. The same red dye is found in the roots of this plant as in those of its near relative, *Galium verum* (Lady's Bedstraw), and if Goose Grass is fed to birds it will stain their bones pink (harmlessly).

This plant, if boiled in broth, was said by Culpeper 'to keep them lean and lank that are apt to grow fat'. He also said that, if boiled in hog's grease, the plant would help 'kernels in the throat' or inflamed tonsils.

USES To make 'Cleavers coffee': in late summer, gather as many of the sticky, clinging seeds (away from any source of pollution) as you have the patience for. Remove bits of stalk, leaves, feathers, animal hairs, etc. spread them out on baking trays in the oven, and roast *very* slowly. When crisp, crush them in a mortar and sieve out the husks. Use as coffee.

The dried plant may be taken as an infusion for insomnia last thing at night, flavoured with lemon and sweetened with honey (1 oz (30 g) to 1 pt (0.6 l) water).

CONTROL The matted growths of this plant are very easily removed by hand and composted.

CAPRIFOLIACEAE Honeysuckle Family

Elder, Bore-wood
Sambucus nigra L.

A Shrub or small tree
AH Up to 10 m (32 ft)
L Leaflets ovate-lanceolate, acuminate, serrate
F Flat-topped cyme. June to July
H Woods, hedges and waste places, nitrogen-rich soil

The name of this well-known shrub comes from the Greek 'sambuca', which means a stringed musical instrument, and skilled country people still make flutes, pipes and whistles from the hollow branches.

There are many legends and superstitions about Elder trees. One legend has it that on Christmas Eve the pith of the larger branches, cut into flat discs, soaked in oil and then lighted and floated in a glass of water, will reveal by its light all the witches and sorcerers of the neighbourhood.

Many damning things are said about Elder wood; it is unlucky to burn it, and it must never be used to build any part of a boat. When hearses were horse-drawn, the drivers liked a handle of Elder for their whips, to protect them against the new spirits of the passengers they carried. No furniture for the house should ever be made of it, least of all a baby's cradle, for evil will befall the child who sleeps in it.

Elder trees are rank in their growth and they smell, and one old country name for them is 'God's Stinking Tree'; another is 'Judas Tree' since Judas Iscariot is said to have hanged himself from an Elder tree. The fungus 'Jew's Ear' (*Hirneola auricula-Judae*) may often be found growing out of the bark as if to prove the legend, whereas common sense will refute it, simply because this variety of Elder is neither tall enough nor strong enough to hang a man comfortably.

In the Tyrol, (where people still lift their hats to the Elder tree), crosses of the wood are made and planted on graves. If the cross bursts into leaf and blossom, the beatitude of the deceased is acknowledged; if the cross remains leafless, the bereaved will know that the deceased lived and died without merit. Another death-superstition was that he who grew an Elder tree in his garden would die in his own house.

USES *Elderflower water* Put 8 oz (225 g) Elderflowers into a stone or earthenware jar and add 1 pt (0.6 l) boiling water. Cover tightly, stand the jar in a saucepan

Elder

of boiling water and simmer for 3 hours. Strain when cool and pour the liquid into a clean glass bottle that has been rinsed round with eau-de-cologne. Cork with a natural cork, and use within 10 days. This is a gentle astringent.

Elderflower cream Slowly melt 16 oz (455 g) pure lard in a saucepan and add as much Elder blossom as the lard will cover (do not leave the flowers on the stalks). Simmer gently for an hour – strain twice while still warm enough to pour, and add a few drops of any pure flower oil (Rose, Carnation, Violet, Wallflower, Heliotrope, Pink, etc.). Pour off into small pots with lids. This is very good for rough skins.

Tonic bath Put Elderflowers in a muslin bag and drop this in the water for a pleasant, refreshing bath.

Make different fritters by picking large clusters of the flowers – wash them well, hold by the stalk, dip them into prepared pancake batter, and fry. Cut off the stem and serve with sugar. Eat them with a knife and fork, as they can be a little untidy, and do not over-indulge – these pancakes have the same effect as unripe gooseberries!

The Elder tree was once called the 'medicine chest of the country people' and all parts of the tree can be used.

The fresh leaves may be made into an ointment for chilblains, swellings, bruises and wounds, as follows: take 8 oz (225 g) Elder leaves, 4 oz (115 g) Plantain leaves, 2 oz (55 g) Ground Ivy and 4 oz (115 g) Wormwood leaves; mince them keeping all the juices. Melt 4 lb (1.8 kg) pure lard very slowly, and stir the herbs and juices into it until the chopped leaves become crisp with gentle cooking. Strain and put into pots.

The fresh leaves crushed up with either lard or olive oil will relieve the pain of haemorrhoids.

The old-fashioned Elderflower tea, which was taken at the onset of a cold, chill or influenza, is still a good panacea. This simple remedy is made as follows: put a handful each of Elderblossom (fresh or dried) and Peppermint leaves into a large earthenware jug and pour over 1½ pt (0.9 l) boiling water. Cover and allow to infuse in the warm for half an hour. Strain and sweeten with honey. Drink as much of this beverage as hot as possible last thing before sleeping.

CONTROL As soon as the young tree is recognized, root it out. Nothing at all will grow in the shade of an Elder tree, but if you have an old tree in a corner, the ground beneath it will make a good place for a compost-heap, as the Elder roots excrete a substance which aids the necessary fermentation in the heap. The tree will grow very quickly so remove it from ornamental walls or the corner behind the garage.

COMPOSITAE Daisy Family

Canadian Fleabane, Horseweed

Conyza canadensis (L.) Cronq. *(Erigeron canadensis L.)*

A   Stiffly erect and fluffy looking
AH 8–100 cm (3–40 in)
L   Narrowly lanceolate or linear
F   Numerous, whitish. August to September
H   Waste ground
    Annual

Another late-summer developer, almost unnoticed with its no-colour flowers, this plant takes on a fluffy appearance when the 'clocks' develop. It came to Europe from North America about 250 years ago and likes the limey rubble at the base of walls and the edges of car-parks.

CONTROL It is very easily pulled up by hand. Try to recognize it by its spiky-leaved appearance before the minute flowers turn into tiny 'clocks', or it will be all over the garden in the following year.

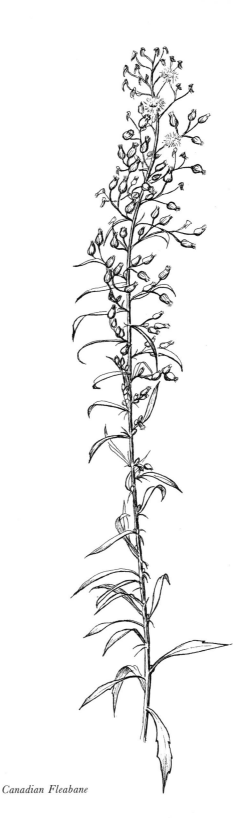

Canadian Fleabane

93

Daisy, Dog Daisy, Day's Eye, Herb Margaret, Bruisewort

Bellis perennis L.

A Rosette-shaped
AH Low
L Obovate-spathulate, rounded at ends, sparsely hairy
R Fibrous
F White with pink undersides. March to October
H Short grassland
Perennial

'When you can put your foot on seven daisies, summer is come' – and what country child has not made a daisy-chain in the summer. One of the Daisy's other names is Herb Margaret, because it was the emblem of Queen Margaret, wife to King Henry VI of England. Another old name for the plant was Bruisewort because the Crusaders made an ointment from the plant to heal their bruises, broken bones and wounds. An old tradition says that if the roots are boiled in milk, and the milk then given to puppies, they will stop growing.

Gerard recommended the plant as a catarrh cure, and said in his earliest herbal – 'The juice of the leaves and rootes snift up into nostrils purgeth the head mightilie of foule and filthy slimie houmours: and helpeth the Megrim.'

USES For a plant surrounded by so much sentiment, the Daisy has surprisingly many medicinal virtues. It is good for heavy menstruation, in the form of a decoction: take 2 oz (55 g) flower-leaves to 2 pt (1.14 l) water, bring to the boil and simmer for a few minutes only. Allow to infuse (covered) for 15 minutes in a warm place and drink a cupful (strained) 3 times a day between meals. This decoction is also reputed to be good for rheumatism, bronchitis and similar ailments; chewing the fresh leaves will help to cure mouth ulcers.

CONTROL If your lawn is a flat mass of Daisy plants, then the soil is deficient in lime, which the Daisy manufactures for itself wherever it grows, and you will have to get down and dig all the plants out with a plantain-lifter, filling up the small holes with a mixture of soil and grass-seed.

Yarrow, Milfoil, Thousand-leaf, Bloodwort, Soldier's Woundwort, Nosebleed Weed, Devil's Nettle

Achillea millefolium L.

A Feathery-leaved
AH 8–55 cm (3–22 in)
L Lanceolate outline, 2–3 times pinnate, dark green
R Stoloniferous
F In dense terminal corymbs, white, sometimes pink. June to October
H Meadows, roadside verges, waste places
Perennial

Yarrow has been a magical and medical herb since the time of Achilles, who is said to have applied the herb to the wounds of his soldiers as they lay bleeding on the battlefield, and hence given the plant its name. The Yarrow has many other old names that encourage this belief, such as Knight's Milfoil, Soldier's Woundwort, Nose Bleed, Bloodwort, Herb Militaris, Sanguinary, Staunchweed and Staunch-grass. A schoolboy's way of leaving school with precipitation was to stuff Yarrow leaves up his nose in order to make it bleed. However, this action can work either way – it can also stop copious bleeding if twisted up and applied into the nostrils (rather like cobwebs).

Yarrow contains an intoxicating constituent and is still used in Scandinavia in the brewing of beer.

A good companion in the herb bed, Yarrow strengthens its neighbours, making them disease-resistant, and intensifying the strength and yield of the aromatic oils from the culinary herbs. It is also a good plant to have in the lawn, since the copper, phosphates and nitrates which it manufactures greatly benefit the neighbouring grasses.

USES Externally, a strong decoction of the fresh leaves (found all the year round) can be used to soothe sore nipples caused by breast-feeding. Yarrow tea is a good old-fashioned remedy for bad colds, and is made as follows: 1 oz (30 g) leaves to 1 pt (0.6 l) boiling water; allow to infuse. Dose: a wineglassful 3 times a day, warm or hot, flavoured with lemon juice. This can also be helpful during children's measles.

Yarrow is used by homoeopathic practitioners to restore menstrual flow when its interruption is caused by an emotional disturbance, without any risk to the

Yarrow

foetus should the condition be an unsuspected pregnancy. A decoction for this condition is made as follows: 2 oz (55 g) flowers and leaves to 2 pt (1.14 l) cold water, bring to the boil and leave to infuse, covered, in a warm place for 15 minutes. Strain. Dose: 3 cupfuls per day. This may also be used as a lotion for treating cuts, suppurating sores and small wounds, and for improving spotty and scabby complexions.

CONTROL Dig up plants and put them in the herb bed, but watch that they do not spread. The plants make excellent compost.

Scentless Mayweed

Daisy

COMPOSITAE Daisy Family

Scentless Mayweed, Scentless Chamomile
Tripleurospermum inodorum (L.) Koch *(Matricaria inodora L.)*

A   Conspicuously large flowers on inconspicuous plant
AH 15–60 cm (6–24 in)
L   Hair-like, pale green
F   White. May to September

H   Roadsides and cultivated land
    Annual

This large-flowered roadside daisy is very similar to the variety that is found on shingle beaches, rocks and walls by the sea (*Matricaria maritima*). The shining, white petals are most attractive when first seen in late spring but by autumn the flowers sprawl and straggle. This plant is related to the true Chamomile (*Anthemis nobilis*) which may be made into a sweet-smelling lawn or used in paths in a herb garden.

CONTROL   Easily rooted up by hand and composted as soon as the young foliage is recognized.

COMPOSITAE Daisy Family

Feverfew, Featherfew, Bachelor's Buttons
Chrysanthemum parthenium L. (Bernh.)

A   Bushy
AH 25–60 cm (10–24 in)
L   Light green, aromatic, narrowly ovate pinnatifid
F   Lax corymbs, white. July to September
H   Waste places, walls and hedgerows
    Short-lived perennial

The name Feverfew is a corruption of the word 'Fevrifuge' which means 'to drive away fever' and this plant was cultivated in the earliest monastery gardens for its fever-dispelling and tonic properties. It was a herb of good luck and, planted near houses, was believed to purify the air and keep the inmates free from disease.

Feverfew

97

Gerard said of the plant that 'it joyeth to grow among rubbish' and this is certainly true – minute seedlings may often be found trying to grow in extremely inhospitable situations. Gerard evidently believed firmly in the virtues of the plant, because he continued 'and two drams of it taken with hony or sweet wine, purgeth by seige melancholy the flegme; wherefore it is very good for them that are giddie in the head, or which the turning called *Vertigo*, that is, a swimming and turning in the head. Also it is good for such as be melancholike, sad, pensive, and without speech.'

The plant is extremely attractive with its mass of yellow-centred, bright-white flowers on a neat bush of yellow-green leaves (these remain throughout the winter). It can be planted under shrubs or trees, and will tolerate extremes of heat and dryness or even the sour dampness against a north-facing wall. However, adult plants resent being moved and will need a great deal of watering and a shady position. The seedlings should be transplanted while still small.

USES The plant contains camphor (bees dislike it) and it will bring immediate relief from mosquito bites if applied fresh to the swellings. It was used as a medieval headache cure, and an infusion of the herb is good for flatulence, indigestion and colds. The infusion is made of 1 heaped tbsp herb to 1 cup boiling water. Dose: 1 or 2 cups a day, 1 tbsp at a time.

Dried branches of the plant laid under paper in drawers will keep moths away.

CONTROL Very easily rooted up and composted, but it will probably reappear for several years if it was long-established.

LILIACEAE Lily Family

Wild Garlic, Ramsons
Allium ursinum L.

A   Ground-covering green leaves, strongly aromatic
AH Up to 30 cm (12 in)
L   Ovate-elliptic, veined, deep green
R   Bulb
F   6–20 flowered, flat-topped. April to June
H   Damp woods, shady places
    Perennial

Wild Garlic smells of garlic from a distance even when quite undisturbed, and the fragile-looking white star-flowers are not for picking unless you have no sense of smell.

USES The leaves will flavour food – the powerful odour vanishes with cooking to leave just a delicate hint of the characteristic garlic savour.

CONTROL It is easy to dig up the bulbs.

Wild Garlic

Pink-flowered and Red-flowered species

Field Poppy

PAPAVERACEAE Poppy Family

Field Poppy, Corn Rose, Headache, Pepper Box, Flanders Poppy, Canker, Redfield
Papaver rhoeas L.

A   Erect and branched
AH 20–60 cm (8–24 in)
L   Pinnately cut and divided, stiffly hairy
R   Slender tap-root
F   4 petals, scarlet, often with blackish-purple blotch at base. June until frosts
H   Cornfields, roadsides
    Annual

The seeds of poppies can lie dormant for years before suddenly germinating to beautify the devastation often worked by man in the name of progress. Legend has it that the poppies which sprang up all over the battle-fields of Flanders grew from the blood of the slain soldiers, so that the Field Poppy became the emblem of Remembrance Day.

USES The Field Poppy is often cultivated for its seeds, which can be scattered on bread and cakes.

The fresh petals were formerly made into a syrup and used as an anodyne. They are still used today for coughs, colds, bronchial complaints, whooping-cough, asthma and insomnia: make an infusion with a handful of fresh or dried petals to 2 pt (1.14 l) boiling water. Allow to cool. Dose: 1 cupful 4 times a day.

For a bath that is different, mix equal quantities by weight of Field Poppy, Periwinkle, True Valerian, Pansy, Male Fern and Maidenhair Fern. Take 2 pt (1.14 l) cold water and add 1 oz (30 g) of this mixture, bring to boiling-point, allow to simmer for 2 or 3 minutes. Turn off the heat and leave to cool. Strain and add the liquid to the warm bath water. This is supposed to have the same effect as rhinoceros horn!

CONTROL Easily tweaked out by hand, but catch it before it seeds.

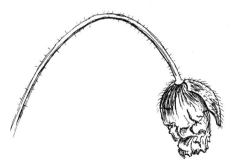

PAPAVERACEAE Poppy Family

Opium Poppy, Mawseed
Papaver somniferum L.

A   Tall and erect
AH 30–120 cm (12–48 in)
L   Glaucous, undulate, ovate-oblong, usually pinnately lobed
F   Petals white, pale pink or lilac, with or without basal purple-black blotch. June to August
H   Waste ground, a garden escape
    Annual
    Poisonous

The true Opium Poppy is very rarely found in the West; what usually appears in the middle of the rose-bed without an invitation is the pale variety of the sub-species *Hortense* which is very similar and which is often grown in gardens. They look well in the mass, and will grow in poor soil as long as the position is sunny, but they do not transplant easily, so if you want Opium Poppies, sow the year before in the place where they are to grow, and thin out accordingly. All parts of the Opium Poppy are very poisonous, but particularly the unripe seed-heads.

In those areas of the world where the plant is grown for its medicinal value, that is, Asia Minor, Turkey, Iran, India and China, the crop is worth a great deal of money. The climate in the West is not generally conducive to the cultivation of Opium Poppies for medicinal purposes, though it was tried in the eighteenth century, and more recently in France and Germany. The main extract from the plant is morphine; other extracts are codeine, narcotine, papaverine, and the narcotic hallucinatory drug, heroin.

The principal ingredient in some cough medicines is a syrup which is prepared from the seed-capsules. This linctus is most effective and pleasant tasting, but great care should be taken not to leave the bottle where children can reach it.

Opium Poppy.

The dried seeds are quite free from any narcotic constituent, and are made into an oil, which is used for cooking and by artists' colourmen.

CONTROL Easily weeded out by hand.

FUMARIACEAE Fumitory Family

Fumitory, Earth-smoke, Wax Dolls, Beggary
Fumaria officinalis L.

A   Weak-stemmed and branching
AH About 30 cm (12 in)
L   Glaucous, leaf segments with flat lanceolate lobes
F   Pink tube with darker tips and wings. May to September
H   Edges of cornfields
    Annual

The name Earth-smoke derives from an early legend that this plant originated in vapour rising from the earth, and in spite of its delicate appearance, it is quite capable of strangling whole cornfields, as the name Beggary indicates.

It has been used medicinally from earliest times as a treatment for arthritis, liver disorders and gallstones; as a diuretic, a laxative, a tonic and to improve sluggish digestions; and as an infusion used externally in the treatment of scabies and eczema.

USES A lotion for clearing the skin may be made by boiling 1 oz (30 g) plant in ½ pt (0.3 l) milk for a few minutes; leave to cool for a quarter of an hour and strain. An infusion of the leaves will clear the skin of unwanted freckles, or banish the last of a summer tan.

The whole plant may be used to dye wool yellow.

CONTROL Hand-weed, except where the growth of the plant is considerable, in which case various chemical weedkillers are effective.

MALVACEAE Mallow Family

Common Mallow, Fairy Cheeses
Malva sylvestris L.

A   Erect and bushy or decumbent main stem
AH 45–90 cm (18–36 in)
L   Roundish, shallow crenate lobes, cordate at base, folded, sparsely hairy

F In axillary clusters, rose-purple with darker stripes. June to September

H Roadsides, waste places, walls
Perennial

The Common Mallow thrives in roadside dust, but it will like the rich soil of your garden even better. The flowers are handsome at first, but as the summer progresses the blossoms get smaller and paler and the plant begins to look ragged and dusty. The flowers used to be woven into May-Day garlands, and the many uses of the Mallow plant have been known since the dawn of civilization. It was made into a soup in ancient Egypt, and Horace said that eating it 'developed the intellectual faculties and encouraged the practice of virtue', while Pliny declared that 'any person taking a spoonful of common Mallow will that day be spared from all maladies that might come his way'.

USES The flowers and leaves have a very high mucilage content, which is made into a salve for skin and eyes. An infusion may be taken internally for stomach complaints, coughs, bronchitis, hoarseness, inflammation of the larynx and tonsils, laryngitis, emphysema, asthma and constipation: add a handful of flowers and leaves to 2 pt (1.14 l) cold water, and soak for 10 minutes. Heat up to boiling point, but do not allow to boil. Remove from heat and allow to infuse for 15 minutes; drink as a tea. (For a laxative, use $\frac{1}{3}$ oz (10g) plant to a cupful of water.)

For toothache, gingivitis, and sore or painful gums, chew some flowers that have been softened in hot water. As a remedy for bee- or wasp-stings, crush some leaves in olive-oil and apply as a poultice.

CONTROL Dig up the young plants before their roots develop – the roots of an adult plant will penetrate walls and paths and wind among hedging plants. Cutting off the top growth alone will merely give the plant a new lease of life, but if it is smothering something special, cut off the aerial parts and touch the exposed root with a proprietary weedkiller.

GERANIACEAE Geranium Family

Cut-leaved Cranesbill
Geranium dissectum L.

A Branched and straggling, hairy
AH 10–60 cm (4–24 in)
L Orbicular, deeply divided into 5–7 lobes
F Reddish-pink, bifid. May to August
H Cultivated ground, hedges, waste land
Annual

This plant is only distantly related to the scarlet bedding-out plant of patio and public park. The wild

geraniums are called cranesbill or storksbill because of the resemblance between the fruit and the beak of the bird. (The Greek for 'crane' is *Geranus*, hence the generic name.) This species has the most finely cut-up leaves of the whole family, and is therefore easily recognized.

CONTROL The plants are very shallow-rooting and are easily pulled up.

GERANIACEAE Geranium Family

Dovesfoot Cranesbill
Geranium molle L.

A Branched from the base and softly hairy
AH 10–40 cm (4–16 in)
L Reniform in outline, irregularly 5–9 lobed
F Numerous, small, bright rosy-purple. April to September
H Dry grassland, waste places and cultivated land
Annual

This is an undistinguished, furry member of the huge Cranesbill family, all of which are easily recognized by the pointed, beak-like seed-pod. Gerard knew of this

Fumitory

Common Mallow

Cut-leaved Cranesbill

104

Dovesfoot Cranesbill

Herb Robert

105

plant, and recommended it as a cure for ruptures, and 'if the ruptures be in aged persons, it shall be needfull to adde thereto the powder of red snailes (those without shels) dried in an oven in number nine, which fortifieth the herbes in such sort'. He went on to say that this mixture never failed to cure 'though the rupture be great and of long continuance' – we now know that ruptures can only be cured by surgery.

The Dovesfoot Cranesbill is one of three ingredients of an old hair tonic which was guaranteed to make hair grow on the baldest head; the other two were Lady's Smock (*Cardamine pratensis*) and Twayblade (*Listera ovata*), though in what proportions is not now known.

CONTROL Easily weeded out by hand, as the root is very small.

GERANIACEAE Geranium Family

Herb Robert, Wren Flower, Robin-i'-th'-hedge, Robin's Eye, Stinker Bobs, Death-come-quickly

Geranium robertianum L.

A   Branched and leafy
AH 10–50 cm (4–20 in)
L   Bright or dark green, polygonal with 5 leaflets, deeply pinnatisect
F   Bright pink. May to September
H   Woods, hedgebanks, rocks and shingle
    Annual or biennial

This well-known plant has many more country names than are given here. It takes its main name from the French Abbé Robert, who founded the Cistercian Order in the eleventh century. However, it is not necessarily a plant with holy associations. The name 'Robin' or 'Robert' was often linked with Robin Goodfellow, a mischievous medieval folk-figure who haunted woodland dwellings, or with the Robin red-breast, a bird believed to bring bad luck, illness or even death if it flew into the house.

In fact, the innocent-sounding name of Robin was usually associated with goblins and evil and all 'Robin' flowers have connotations of death, the Devil, fairies, snakes, sex and cuckoos.

USES The leaves were used in medicine from earliest times, and were crushed to make compresses to heal external bruises and wounds. A handful of well-squashed leaves, rubbed over exposed skin areas, will keep mosquitoes away. The plant may be used to make an infusion for all forms of haemorrhage, kidney and bladder ailments and, formerly, for tuberculosis: take equal parts of flowers and leaves to water, and chop the plant finely to release the juices. Place all in an enamel or earthenware vessel, and heat gently until boiling point is nearly reached. Keep over the heat for a few minutes, but do not allow to boil. Remove from heat and let steep overnight. Strain. Dose: half a cup morning and evening. (Use double the quantity of plant to water to make a ringworm remedy and to kill lice and their eggs.)

CONTROL Very easily weeded out by hand but catch it before the seeds are dispersed.

BALSAMINACEAE Balsam Family

Himalayan Balsam, Policeman's Helmet, Jumping Jack

Impatiens glandulifera Royle

A   Erect and robust
AH 1–2 m (3–6 ft)
L   Dark green, opposite or in whorls of 3, lanceolate, sharply serrate
F   Large, purplish-pink and helmet-shaped. July to October
H   Ditches and river-banks
    Annual

This handsome plant likes to have wet feet, and plants established on dry soil are not so tall or well-formed. The ripe, green, seed-pods fascinate children as, at a touch, they zip open faster than the eye can see, and the three parts of the seed-case coil up like a trio of corkscrews, scattering the seeds widely in the process. As its name implies, the plant comes from the Himalayas, where it grows to a height of over 10 ft (3 m) but it has now naturalized most successfully throughout Europe.

If you have a pond, it is worth planting some seedlings in the mud at the water's edge, but you will have to 'capture' the ripe seed with a carefully inverted paper-bag before the capsule explodes.

CONTROL Easily removed by hand-weeding.

PAPILIONACEAE Pea Family

Alsike Clover
Trifolium hybridum L.

A Erect or decumbent
AH Up to 60 cm (24 in)
L Leaflets serrate, obovate or elliptic
F Globular heads, pink and white. June to August
H Roadsides and waste places
   Perennial

This clover is included because of its differences from other clovers – the flower-heads are bigger and bolder than its nearest relative, the white Dutch Clover, which has different leaves and a more creeping habit. The Alsike Clover is a handsome, upstanding plant, and, like all clovers, its roots manufacture great stores of nitrogen. The vivid green leaves are very good in the compost-heap, and the remaining roots will enrich the soil.

CONTROL Cut off the tops and compost them, or dig them in as green manure.

Alsike Clover

PAPILIONACEAE Pea Family

Red Clover, Purple Clover, Bee Bread, Honeysuckles, Clovers
Trifolium pratense L.

A Erect and bushy
AH Up to 60 cm (24 in)
L Leaflets elliptic to obovate, with whitish crescentic spot
F Globose, pink-purple. May to September
H Grassy places
   Perennial

This variety called Red Clover is more often a dull pink or even mauve-purple and it may be recognized from a distance because it often grows in a large solitary clump, almost a bush. The clovers were the 'clavers' of the Middle Ages, and several English place-names are derived from areas where the plant grew well: Claverton, Cheshire, Clavering, Essex and Claverdon, Warwickshire.

It is the food plant of the Narrow-bordered Five-spot Burnet moth, *Zygaena lonicerae* L.

USES The flowers and leaves, made into a tea, traditionally are a gentle remedy for flatulence, for soothing the nerves, and for restoring fertility, and also an aid for children with whooping-cough: take 1 oz (30 g) fresh flowers and leaves, and pour 1 pt (0.6 l) boiling water over them. Allow to cool, keeping covered. Flavour with lemon and sweeten with honey if liked. Dose: a wineglassful several times a day.

A fomentation of the leaves and flowers may be applied externally to soften painfully hard milk glands, and athlete's foot sometimes responds to a poultice of the crushed flowers and leaves.

CONTROL Easily pulled out of the ground and excellent for the compost-heap because of the high nitrogen content that all clovers have. If you have acres of it and need to use a chemical weedkiller, the dead plants cannot, of course, be composted; an alternative is to dig in the leaves as green manure.

Himalayan Balsam

Red Clover

Common Vetch

PAPILIONACEAE Pea Family

Common Vetch, Fitch, Tare
Vicia sativa L.

A Trailing and tendrilled
AH 15–150 cm (6–60 in)
L Paired leaflets, linear to obovate, tendrils simple
    or branched
F Cerise. May to September
H Hedges and grassy places
    Annual

This vetch was formerly grown as a fodder crop for cattle, but has now naturalized itself all over northern Europe.

CONTROL As for other vetches (see p. 138).

Blackberry

ROSACEAE Rose Family

Blackberry, Bramble, Briar
Rubus ulmifolius (Schott)

A Bush with rooting, arching stems
AH Up to 3 m (10 ft)
L Leaflets 3–5, dark green, glabrescent above, white beneath
F Petals bright purplish-pink, crumpled. July to August
H Woods, scrub, hedges and heaths

This is a plant which has flowers and fruit on the same bush at the same time. The prickles vary in length, strength and spininess.

Like the Stinging Nettle, the Bramble bush needs no description – it has been known to man since his beginning. Blackberry pips have been found in the stomach of Neolithic man and Brambles are mentioned in the Bible in the parable of the trees who tried to choose a king among them. After the Vine, the Fig, and the Olive had refused the honour, the Bramble was elected (Judges 9:8–15).

Brambles are the food plant of the Broom moth (*Ceramica pisi* L.), the Emperor moth (*Saturnia pavonia* L.), the Fox moth (*Macrothylacia rubi* L.), the Great Brocade moth (*Eurois occulta* L.), the Green Hairstreak butterfly (*Calliophrys rubi* L.) and the Grizzled Skipper butterfly (*Pyrgus malvae* L.).

USES The Blackberry bush has many virtues, the most well-known being the ripe, succulent berries in late summer. These are delicious eaten raw (washed and picked over), with cream and sugar, or baked with apple in a pie. The berries should be picked before Michaelmas (29 September), after which they do not taste good. Legend has it that the Devil defiles them all on Michaelmas Eve by spitting on them – or worse.

An excellent Blackberry wine may be made as follows: pick a large quantity of fruit and crush it in a preserving pan, using a large round plate. Add 2 pt (1.14 l) water to 1 gal (4.5 l) crushed fruit, and allow to stand, stirring occasionally, until the next day, keeping covered. Strain off the liquid. Add 2 lb (906 g) sugar to 1 gal (4.5 l) liquid; put into a wooden cask and cork tightly; leave until the following October.

The plant also has medicinal value. The bark of the roots contains tannin, as do the leaves to a lesser degree, and the root-bark makes an excellent astringent and remedy for diarrhoea. Peel the bark off the root

and dry thoroughly, turning often. Add 1 oz (30 g) grated root to 1½ pt (0.9 l) water, bring to the boil and reduce to 1 pt (0.6 l). Dose: 4 fl oz (115 ml) every 2 hours.

Gather the leaves in the summer, dry them, and use 1 oz (30 g) to 1 pt (0.6 l) water. This decoction, with honey added, is remedial for sore throats, mouth ulcers, sore gums, and as a gargle and mouthwash. A poultice of crushed leaves is good for burns and the swellings of insect bites; the fresh leaves, crushed up, will arrest bleeding caused by the thorns when picking the fruit.

Before the berries are fully ripened, they may be picked to make a syrup with an equal weight of sugar. One tbsp syrup 3 times a day eases catarrh and sore throats.

CONTROL Tweak up the small plants as soon as recognized. If there are large areas of old Briars to be cleared, use a new bill-hook, and wear heavy gloves and protective clothing when cutting the growth down. Burn the cut lengths. The ground will have to be rotovated to remove the roots. Note that flame guns are not really suitable to clear Blackberries, as they tend to char the leaves but leave behind a forest of barbed stems. If there are only one or two elderly Brambles with inaccessible roots, cut off all the top growth and tie a pad soaked in sodium chlorate over the cut ends.

ONAGRACEAE Willowherb Family

Broad-leaved Willowherb
Epilobium montanum L.

A   Tall and erect
AH 20–60 cm (8–24 in)
L   Usually opposite, ovate-lanceolate, sharply and irregularly toothed
R   Stoloniferous
F   Pale rose, deeply notched. June to August
H   Woods, hedges, walls, rocks and gardens
     Perennial

This plant is unexciting in form and flower, and is all-too-common in gardens. The plants are found in scattered clumps, in the most inhospitable places, and seem to grow invisibly until they reach maturity, by which time they are seeding themselves all over the garden.

CONTROL Determined searching out of parent plants before they seed is the most satisfactory method of extirpation. For a short time the garden will be clear, but those irritating little leaf-rosettes will spring up again, and the performance will have to be repeated several times.

ONAGRACEAE Willowherb Family

Rosebay Willowherb, Fireweed, Blooming Sally
Epilobium angustifolium L.

A   Tall and vigorous
AH 30–120 cm (12–48 in)
L   Spirally arranged, numerous, narrowly oblong-lanceolate, waved at the margins, conspicuously veined
R   Horizontally spreading
F   Petals rose-purple. July to September
H   Rocky places, wood clearings, burnt areas, railway embankments
     Perennial

Fireweed is the apt name for this plant because the swathes of brilliant magenta flowers are the first to repopulate burned woodland or a blackened heath. The plant was a scarcity in Gerard's time, and was grown as a rare garden flower, whose colour he described as 'Orient purple'. It now springs up in huge numbers as if by magic when centuries-old buildings are demolished. This Willowherb does not spread by magic, however, but makes doubly sure of its own survival by regenerating through a system of creeping stems and by producing a vast quantity of lighter-than-air plumed seeds.

The leaves of the plant (called Willowherb because of its resemblance to the Willow tree) were formerly used as a substitute for, and additive to, tea, and are still used for this purpose in Russia, the beverage being called 'Kaporie tea'. The plant grows in glorious profusion near Seattle, in America, and the bee-keepers of the area are able to market a speciality called 'Firewood honey'.

Medicinally, the plant is not generally used now, but in the past it was used in the treatment of whooping-cough and asthma, and in America it is still used as an intestinal astringent.

CONTROL If you have a very large garden, a splash of magenta Willowherb against the blue-green of pines is

Broad-leaved Willowherb   *Rosebay Willowherb*

113

Enchanter's Nightshade

a lovely sight. However, few people can allow this, and to dispose of the plant, the strong running roots must be burned, and all the plant removed before it goes to seed. Even then, new plants will probably grow from seeds floating in from a distant patch.

ONAGRACEAE Willowherb Family

Enchanter's Nightshade, Philtrewort, Witch-flower
Circaea lutetiana L.

A   Slender and upright
AH   20–70 cm (8–28 in)
L   Ovate, gradually acuminate, sinuate-toothed
R   Semi-stoloniferous
F   Small, pale pink. June to July
H   Woods and shady places
     Perennial

This plant, despite its beautiful names, is insignificant looking. It is not a member of the Nightshade family (*Solanaceae*) but is related to the Fuchsia, Clarkia and Godetia. In Germany the plant was formerly linked with witchcraft and called *Hexenkraut* (Witches' Cabbage), but there appears to be no record of its magical use elsewhere.

CONTROL Easily rooted out once identified. It has semi-stoloniferous roots, so care should be taken to uncover and remove all the pieces. The hooked seeds will catch on your clothes and be distributed to other parts of the garden so try to remove the plant before the seeds form.

POLYGONACEAE Dock Family

Common Knotgrass, Matgrass, Doorweed, Pinkweed, Birdgrass, Stonegrass
Polygonum aviculare L.

A   Erect or trailing
AH   3–200 cm (1–80 in)
L   Elliptic, lanceolate, linear
F   Pink. July to October
H   Waste places, arable land and sea shores
     Annual

This small plant grows everywhere, on the edges of dusty roads, in the cracks between paving stones, round the base of gateposts and in lawns. It has ancient medicinal associations.

USES A decoction of Common Knotgrass was a traditional remedy for diarrhoea and dysentery. Take 1 oz (30 g) fresh plant (or double the quantity of dried) to 2 pt (1.14 l) water. Boil for 15 minutes, leave to infuse for about 10 minutes, strain. Dose: half this quantity a day in cupfuls. The same decoction may be used in the treatment of cystitis and is remedial in circulatory disorders such as phlebitis, varicose veins and haemorrhoids.

This is another 'wound herb', being rich in tannin and silicic acid, and the washed plant may be used as a poultice. Always make sure to gather the plant away from roadsides or from farmland that has been sprayed with toxic herbicides.

CONTROL Hand-weed before it seeds. Valuable in the compost-heap.

POLYGONACEAE Dock Family

Persicaria, Redleg, Redshank, Willow Weed, Heartweed, Lady's Thumb
Polygonum persicaria L.

A   Branched and erect
AH 25–75 cm (10–30 in)
L   Lanceolate, ciliate, often dark-blotched
F   Stout, obtuse, pink. June to October
H   Waste places and cultivated land
    Annual

There are several legends about the dark blotches on the leaves of this plant, most of them religious; one tells how the Virgin Mary picked the plant and left the mark of her finger on the leaves. Another, quite common about plants with spotted leaves, is that it grew at the foot of the Cross and was splashed with the blood of Christ.

USES An infusion of Redleg is an astringent and a diuretic (1 tbsp herb to 1 cup water. Dose: 2 cups per day, made fresh). This infusion has also been remedial in the treatment of diarrhoea and jaundice.

Common Knotgrass

CONTROL Easily pulled out of the ground without disturbing other border plants. Good in the compost-heap.

URTICACEAE Nettle Family

Helxine, Mother-of-thousands, Mind-your-own-business
Helxine soleirolii (Req.)

A   Mat-forming
AH Low
L   Suborbicular, subsessile
R   Fibrous
F   Small, pink. May to October

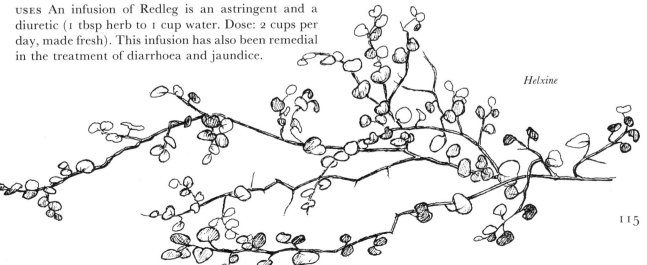

Helxine

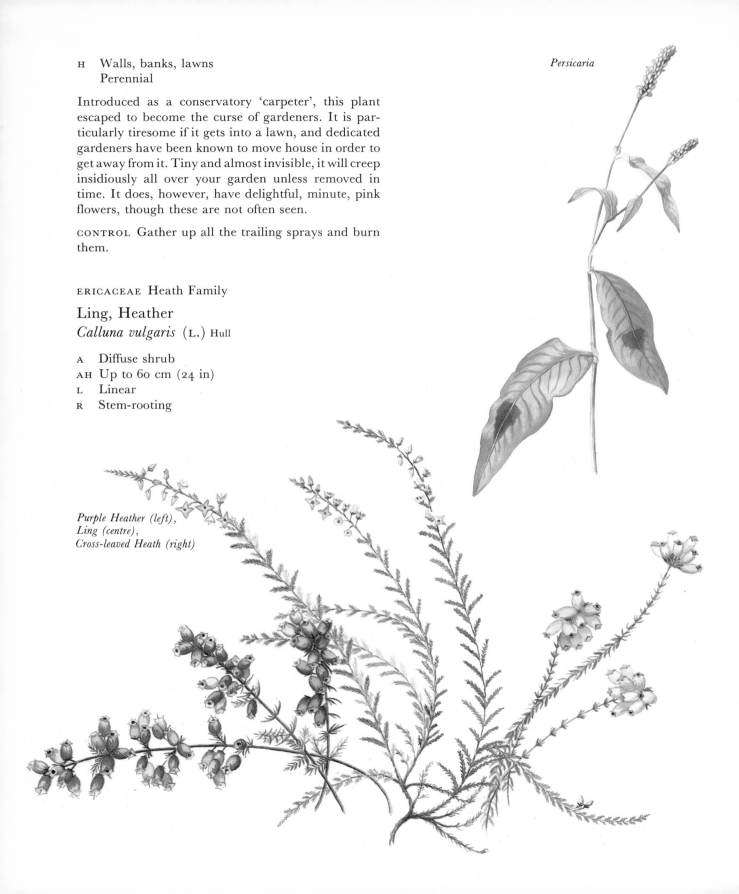

H Walls, banks, lawns
   Perennial

Introduced as a conservatory 'carpeter', this plant escaped to become the curse of gardeners. It is particularly tiresome if it gets into a lawn, and dedicated gardeners have been known to move house in order to get away from it. Tiny and almost invisible, it will creep insidiously all over your garden unless removed in time. It does, however, have delightful, minute, pink flowers, though these are not often seen.

CONTROL Gather up all the trailing sprays and burn them.

ERICACEAE Heath Family

Ling, Heather
Calluna vulgaris (L.) Hull

A Diffuse shrub
AH Up to 60 cm (24 in)
L Linear
R Stem-rooting

Persicaria

Purple Heather (left),
Ling (centre),
Cross-leaved Heath (right)

F   Raceme pale pinkish-purple. July to September
H   Heaths and moors

One legend has it that the name Heather goes back to the time when Kenneth the Conqueror attempted to convert the Picts to Christianity. The Picts fought bitterly and in vain to defend their beliefs, and the plants that were bedewed with 'Heathen' Pictish blood became known as 'Heath' or 'Heather'. The story goes on to say that Kenneth offered to spare the lives of the two remaining tribesmen if they would give him the secret of their famous moorland beer. They refused and the secret of the drink, which was made from Heather, was never divulged, though on the island of Jura, Heather beer is still made.

Heather has had and still has many uses – as thatch for roofs, material for mattresses, brooms (besoms), fuel and food – Heather honey is dark and delicious. It is the food plant of the Emperor moth (*Saturnia pavonia* L.) and the Oak Eggar moth (*Lasiocampa quercus* L.)

Heather grows well only on well-drained soil, and if you have this type of soil, make the most of it to grow the many types of heather which will flower the year round.

USES Put the flowering tips in a muslin bag and add to a bath to help relieve rheumatism and gout.

Make a decoction, beneficial for cystitis, nephritis and rheumatism, of 1 oz (30 g) Heather to 2 pt (1.14 l) water. Reduce by boiling to a third; strain. Dose: drink this quantity during one day.

'Heather Tea', a tonic tea beloved by Robert Burns, is made as follows: take a handful of each of the leaves of Blackberry, Speedwell, Wild Thyme, Wild Strawberry, Heather and Bilberry. Lay these to dry (pick double quantities to last the winter). Roll on a pastry board with a rolling pin to separate hard stems and dried leaves; pick out the stems and mix the remainder thoroughly. Use 1 tsp to a cup of boiling water, and sweeten, preferably with Heather honey.

CONTROL The knotted roots are very determined, so if an area is to be cleared, it is better to cut, burn and rotovate than hand-dig, which is discouragingly slow.

ERICACEAE Heath Family

Cross-leaved Heath
Erica tetralix L.

A   Diffuse shrub
AH   Up to 60 cm (24 in)
L   4 in a whorl, linear
R   Stem-rooting
F   Urceolate, ovoid, rose-pink. July to October
H   Bogs, wet heaths and moors

Not so common in the garden as Ling, but as more areas of moorland are being encroached upon by the property developers, you may find, when the builders depart, that you have a garden-full of *Erica tetralix*, which has been growing there since the last of the great saurians vanished. In former times, London's West-End tailors would dress their windows with this heath when 'the glorious 12th' (12 August, the opening of the grouse-shooting season) was imminent, to remind shoppers to order the correct tweeds for the shooting season.

CONTROL As for Ling.

ERICACEAE Heath Family

Purple Heather, Bell-heather, Scotch Heather
Erica cinerea L.

A   Diffuse shrub
AH   Up to 60 cm (24 in)
L   3 in a whorl, linear, glabrous, dark green
R   Stem-rooting
F   Urceolate, ovoid, crimson-purple. July to September
H   Dry heaths and moors

This is the heather that colours the moors in late summer with rolling waves of crimson-purple – the fluted, red-purple bells are of a richer colour than the other heaths.

CONTROL As for Ling.

Scarlet Pimpernel, Poor Man's Weather-glass

Anagallis arvensis L.

A   Procumbent
AH   6–30 cm (2½–12 in)
L   Ovate-lanceolate, glabrous
F   Light red. June to August
H   Cultivated land, roadsides
     Annual or perennial

This familiar plant with its tiny, light-red flowers was made famous by Baroness Orczy in her novels about a swashbuckling Englishman who, during the French Revolution, called himself 'The Scarlet Pimpernel', and left cards with a picture of the little red flower wherever he went.

The flowers quickly close if the sky becomes overcast or at the onset of wet weather, and they do not open at all if the day is dull and sunless. They do not open even in midsummer until about 8 AM and close in mid-afternoon.

The plant has many magical associations, not least that if you stand and hold it, you will acquire second sight and will hear and understand the language of birds and animals. In former times it was considered sovereign against spells, and would even draw splinters from the flesh if they had been driven in by witches.

The plant has been used by herbalists for centuries, and in the past doctors have used a decoction of it in cases of melancholia and epilepsy. Gerard said that 'the juyce purgeth the head by gargarising or washing the throat therewith; it cures the tooth-ach by being snift up into the nosethrils, especially in the contrary nosethril'. In America, Scarlet Pimpernel is used in very small quantities as a nervine, a stimulant, a diuretic, an expectorant, a diaphoretic and as a cholagogue. However, larger doses cause heavy perspiration and increased kidney activity, and this plant should never be used without advice from a qualified homoeopathic practitioner. The leaves can cause dermatitis to some skins.

CONTROL Very easily weeded out but always wear gloves to handle it.

Field Bindweed

CONVOLVULACEAE Bindweed Family

Field Bindweed, Lesser Bindweed, Wild Morning Glory, Cornbine, Hedge Bindweed, Devil's Vine
Convolvulus arvensis L.

A   Scrambling
AH   20–75 cm (8–30 in)
L   Oblong, ovate, hastate or sagittate
R   Rhizomes stout, spirally twisted, penetrating the earth to a depth of 2 m (6 ft)
F   Funnel-shaped, pink and white striped. June to September
H   Cultivated land, waste places, roadsides and in short turf
     Perennial

Farmers hate this plant because it is a sign of bad husbandry to see the ice-cream striped trumpets twiddling up the cornstalks, and because ploughing will not destroy the roots of *C. arvensis*, which have even been known to grow to a depth of 20 ft (6 m). The large, bell-shaped flowers do not open on dull days, and close quickly at the onset of bad weather.

     The plant is the food of the Convolvulus Hawk moth, *Herse convolvuli* L.

CONTROL Continual pulling out of the long, thready roots is the only way to prevent your garden becoming a plaited mass of Field Bindweed. Everyone suffers from the suffocating stranglehold of this plant, and the better your soil, the finer will be the flowers! The arrow-shaped leaves are very recognizable and can, with the attached portion of root, be hauled out easily, particularly on a damp day.

SCROPHULARIACEAE Figwort Family

Foxglove, Dead Men's Bells, Fairy Thimbles
Digitalis purpurea L.

A   Tall and vigorous
AH   50–150 cm (20–60 in)
L   Ovate-lanceolate, crenate, softly pubescent
F   Raceme of 20–80 tubular flowers, pinkish-purple with deeper purple spots inside on a white ground. June to September
H   Open places in woods and on heaths, acid soils
     Biennial
     Poisonous

Foxglove

The origin of the name Foxglove is charming – it is said that the bad fairies gave the flowers to the foxes to put on their paws so that they could creep silently on to their prey. This has always been a flower associated with magic – if the fairies left a changeling child on your step, having spirited your babe away, Foxglove juice would bring your own child back again.

The plant is famous for its extract, digitalin, which is used to stimulate or regulate the action of the heart. Only the leaves of second-year plants are used, when the flowering spike is partly in bloom, and magenta-flowered plants are preferred to the lighter shades. All parts of the Foxglove are poisonous, and the poison remains active even when the plant is dried. Deaths have been recorded when cattle and horses have eaten hay containing dried Foxglove leaves.

The plant is used medicinally in minute quantities as a diuretic and in cases of dropsy, and many remedies have been described by Culpeper in his herbal. However, these remedies are very dangerous and should *not* be tried in any circumstances.

The plants thrive only in acid soils and though associated with woodland glades, they are quite happy in full sun. The seedlings must be moved into their flowering position when still small; allow plenty of space between plants, as they can grow to 2 m (6 ft) in height.

CONTROL Use gloved hands to pull up the plants as soon as recognized. The poison is not in any way active in the soil and the compost-heap is a safe depository for the plants.

SCROPHULARIACEAE Figwort Family

Red Bartsia, Brownweed
Odontites verna (Bell.) Dum.

A   Pink-tinted, small, stiff and shrubby
AH Up to 50 cm (20 in)
L   Lanceolate
F   Tubular, purplish-pink. June to July
H   Cultivated fields and waste places
    Annual

This is an inconspicuous plant in spite of its reddish tone, which may grow in your hedgebank quite unnoticed. The stem is strong and woody and the plant is stiff to the touch.

CONTROL Easily weeded out with a hearty pull.

Red Bartsia

Scarlet Pimpernel

Hedge Woundwort

LABIATAE Labiate Family

Hedge Woundwort, Wood Woundwort
Stachys sylvatica L.

A Tall and nettle-like
AH 30–100 cm (12–40 in)
L Ovate, coarsely crenate-serrate, hairy
R Rhizomatous
F In whorls, about 6 flowered in a spike. July to August
H Woods and shady places
   Perennial

This sombre-hued flower of shady gardens grows there as a compliment to the richness of the soil. In former days it had a reputation for healing wounds. Gerard relates: 'It chanced that a poore man mowing of Peason did cut his leg with a sithe, wherein hee made a wound to the bones, and withall very large and wide, and also with great effusion of bloud: the poore man crept to this herbe, which he bruised with his hands, and tied a great quantitie of it unto the wound with a piece of his shirt, which presently stanched the bleeding, and ceased the paine, insomuch that the poore man presently went to his daies worke againe, and so did from day to day, without resting one day untill he was perfectly whole; which was accomplished in a few daies.'

CONTROL As soon as this plant is recognized, with its opposite-branching, furry green leaves and wine-coloured flowers, haul it out with all its branching roots and compost it.

LABIATAE Labiate Family

Red Dead-nettle, Purple Dead-nettle
Lamium purpureum L.

A Usually purplish tinted
AH 10–45 cm (4–18 in)
L Ovate, obtuse, regularly crenate-serrate, pubescent
F Tubular, pinkish-purple
H Cultivated ground and waste places
   Annual

Red Dead-nettle varies in appearance according to its location in your garden. On a hot dry bank, or even in a neglected corner of the vegetable garden, it spreads sideways, is no more than 15 cm (6 in) high, and is usually pink or purple-tinged; among the tangle of an overgrown herbaceous border it becomes tall and straggly and stays a uniform green colour.

The leaves, when bruised and used as a poultice, are said to staunch blood flowing from a deep cut.

CONTROL The plant sprawls from a central growing point and is easily grasped. Pull it up and compost it.

VALERIANACEAE Valerian Family

Red Valerian, Padstow Pride, Soldier's Pride, Pretty Betsy
Centranthus ruber (L.) DC.

A Erect
AH 30–80 cm (12–32 in)
L Ovate-lanceolate, glabrous
R Long and penetrating
F Pink, crimson or white. May to August
H Walls, quarries and cliffs
   Perennial

The luxuriant, pink mop-heads of this plant are very attractive. It is a plant of walls and cliffs and should not be confused with the true (more medicinally used) Common Valerian, *Valeriana officinalis*, which grows in ditches and similar damp places.

The plant cannot be moved when it is fully grown, as its fangy roots penetrate very deeply, so if you wish to grow Red Valerian, collect the seeds, sow in a limey mixture (adding some dust swept up from the paths into the seed compost, as it dislikes rich soil), and plant out the seedlings, when they are about 7 cm (3 in) high, into their permanent positions. More seedlings will soon appear if the soil is left undisturbed near the parent plant for the winter period. It likes best to grow on and out of walls, where it is seen to advantage; or at the edge of paths or paving where there is likely to be limestone dust. As it is not a tidy plant, it is better grown in the mass, rather than in separate clumps.

CONTROL The roots can never be pulled out of a wall or paving without major reconstruction, as they are exceedingly long, tough and knotty. If you must clear your walls, paths and patios, keep shearing off the aerial parts, though the plant will take some years to eradicate.

COMPOSITAE Daisy Family

Winter Heliotrope
Petasites fragrans (Vill.)

A   Persistently leafy
AH 25–30 cm (10–12 in)
L   Roundish, deeply cordate, equally serrate
R   Rhizome far-creeping
F   Lilac-pink and vanilla-scented. November to March
H   Stream-sides and waste places
    Perennial

This plant was discovered in Italy in 1800 and introduced to the gardens of wealthy Parisians where it was grown in pots to perfume the winter air. From Paris it was only a short distance to the gardens of the English aristocracy, from whence it escaped to establish itself in scattered colonies all over England.

It flowers at a time when the garden is almost bare, and a few roots may be transferred to tubs or large flowerpots which may be brought indoors in winter, though they will flower better in the arctic conditions of the January garden where, on a mild day, the scent of almonds will be very strong. It must be firmly controlled or it will smother the entire garden, but as it will grow on dry banks under trees it is worth cultivating as ground cover.

CONTROL The thick rhizomes knit themselves into tree-roots and can never be removed without damaging the trees. Keep shearing off the leaves as they grow to control it.

Red Dead-nettle

Red Valerian

124

Winter Heliotrope

Blue-flowered species

Monkshood

RANUNCULACEAE Buttercup Family

Monkshood, Wolf's Bane, Aconite
Aconitum anglicum Stapf

A   Tall, vigorous and leafy
AH 50–100 cm (20–40 in)
L   Pentagonal in outline, palmately 3–5 partite,
    then deeply laciniate, light-green
R   Blackish tuberous tap-root-like stocks
F   Helmet-shaped, blue-purple. May to June
H   Shady stream banks
    Perennial
    Very poisonous

The beautiful but deadly Monkshood is found in old cottage gardens as a border plant, and is very ornamental with its light green leaves and brilliant, blue-purple, hooded flowers. The garden variety is *A. napellus*, which has darker green leaves that are less cut up, and darker blue flowers which come later in the year than *A. anglicum*, the wild plant. The whole plant, but particularly the root, is extremely poisonous and the poison used to be extracted to tip arrows and to poison meat left as bait for wolves, hence its other name of Wolf's Bane. In medieval times poison was used very much more than it is today and every man grew his own, more to dispose of vermin than his neighbour. Monkshood poison was quick and reliable, and the plant therefore found a place, separately from the herbs and vegetables, in old gardens.

Monkshood and Deadly Nightshade were two of the legendary ingredients of the magic salve of witches' 'flying ointment'. Together they would cause delusions and giddiness but the salve also called for Hellebore root, Hemlock, the fat of a murdered babe, soot (as thickener), and bat's blood (as an aid to night-flying!).

Monkshood is much used today as an anodyne, a sedative and a febrifuge, and the preparations obtained from the plant are used to ease the pain of sciatica, arthritis, neuralgia, gout, rheumatism, and certain chronic skin problems. However, Monkshood is more deadly than prussic acid, and acts with more rapidity, and under no circumstances should this plant be experimented with. If it is suspected that any part of the plant has been ingested, a doctor should be called at once.

CONTROL Always use gloves to handle the plant. If you wish to keep it in the garden, move it to the herbaceous border and let it be – it likes semi-shade and humus-rich soil. Otherwise, root it up, and burn it since it will propagate itself from fallen pieces of root and may not compost properly.

BORAGINACEAE Borage Family

Wood Forget-me-not, Remember-me
Myosotis sylvatica Hoffm.

A   Pubescent and branching
AH 15–45 cm (6–18 in)
L   Lanceolate, pubescent
F   Lax cyme of bright blue flowers. March to May
H   Cultivated ground, roadsides
    Perennial

Wood Forget-me-not

This is the flower whose haze of blue sets off the tulips to such advantage. The name comes from a German legend of a knight who walked beside a deep river with his new bride. He stooped to pick for his love some of the bright blue flowers growing by the water's edge (probably the Water Forget-me-not, *M. scorpioides*), but he overbalanced and fell in. As he was swept away by the current, to his death, he tossed the flowers to his love, calling out 'Vergiss mein nicht!' ('Forget me not!')

The plant also used to be called Scorpion Grass because the coiled flower-head resembles the curled tail of the scorpion.

CONTROL If you want to grow more of these pretty spring flowers, leave the parent plants in situ until their mildewed leaves and stems have yielded up the seed harvest. When you finally tweak them out, the progeny will be there in their thousands, only needing to be thinned. Otherwise, pull them up as soon as they flower, before they have a chance to seed.

SCROPHULARIACEAE Figwort Family

Germander Speedwell, Bird's Eye
Veronica chamaedrys L.

A   Prostrate
AH 20–40 cm (8–16 in)
L   Triangular-ovate, coarsely crenate-serrate, dull green, hairy
F   Deep bright blue with white eye. March to July
H   Grassland, woods, hedges
    Perennial

A spikier plant than the Persian Speedwell, *V. persica*, this plant has even bluer flowers which fall off at a touch. Many years ago, the flowers were sewn on to the coats of travellers to protect them from harm and speed them on their way.

CONTROL As this is a perennial, the plant is a little more tenacious of the soil than *V. persica*, but is too small to resist a dedicated tidying up of the border or kitchen garden.

SCROPHULARIACEAE Figwort Family

Persian Speedwell, Buxbaum's Speedwell
Veronica persica Poir

A   Trailing and prostrate
AH Low
L   Triangular-ovate, shortly stalked, coarsely crenate-serrate, light green
F   Bright blue, lower lobe paler. Throughout year
H   Cultivated ground, disturbed soil
    Annual

In the late spring, in a neglected corner of the kitchen garden, the sky-blue flowers of this most common of the Speedwells will suddenly appear. This is an enthusiastic annual, which will proliferate with astonishing speed and vigour.

CONTROL Systematic and thorough weeding must be done before a single seed can form and fall.

VALERIANACEAE Valerian Family

Lamb's Lettuce, Corn Salad
Valerianella locusta (L.) Betcke

A   Slender and erect
AH 7–40 cm (3–16 in)
L   Glabrous, entire, spathulate to sub-acute
F   Cymes, small pale blue flowers. April to June
H   Arable land, hedgebanks, dry soils
    Annual

This plant, because of its tiny, pale-blue flowers, may sometimes be mistaken for forget-me-not. It is a very persistent garden-invader and seems not to mind the mower, but it makes a colourful and edible edging to a path in the vegetable garden. It was called Lamb's Lettuce because it appears at lambing-time.

USES The leaves make a good winter salad, but are best eaten when young, which is from October onwards.

CONTROL Remove the plant before it has a chance to seed. If it is in the lawn, it will have to be painstakingly dug out by hand.

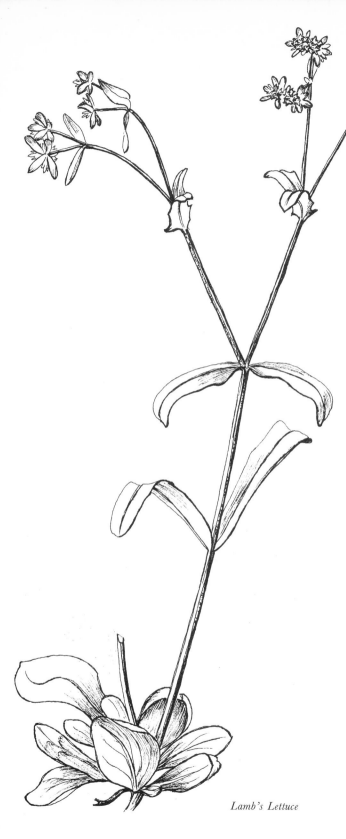

Lamb's Lettuce

Bluebell, Wild Hyacinth, Harebell
Endymion non-scriptus L. *(Hyacinthus non-scriptus)*

A   One-sided spike of blue bells
AH 20–50 cm (8–20 in)
L   Glabrous, linear
R   Bulb
F   Nodding racemes of violet-blue. April to June
H   Woods and hedgebanks
     Poisonous

Bluebell-blue is deceptive – in the mass there seems to be nothing bluer, but looked at individually the flowers are several shades of mauve, violet and purple. The proper name of the flower comes from a Greek legend about the beautiful youth Hyacinthus who was beloved simultaneously by Apollo the Sun-god and Zephyrus the god of the West Wind. Hyacinthus loved Apollo the more, which enraged the jealous Zephyrus, and when Apollo and Hyacinthus were playing quoits, Zephyrus blew the quoit out of its proper course so that it killed Hyacinthus. The grief-stricken Apollo raised from the blood-soaked ground a purple flower with the words 'Ai, Ai' marked upon it, so that his woe would evermore be known. This marked variety of Wild Hyacinth grows in Greece but not northern Europe, where the unmarked variety is called *Endymion* or *Hyacinthus non-scriptus* ('not written-on').

The bulbs are sticky to the touch and it is said that it is possible to use this natural 'glue' as a simple adhesive; it was certainly used to stiffen Elizabethan ruffs.

CONTROL The bulbs grow small bulblets, and the seeds also form tiny bulbs which take about 3 years to develop and flower, so there is a constant succession of new plants. Pick the flowers so that no seed is formed, which will begin to break the cycle, and dig up all the plants that you can see before the leaves disappear.

Germander Speedwell

Persian Speedwell

Bluebell

Mauve-flowered
and Purple-flowered
species

RANUNCULACEAE Buttercup Family

Aquilegia, Columbine, Culverwort, Grannybonnets
Aquilegia vulgaris L.

A Erect and leafy
AH 40–100 cm (16–40 in)
L Glabrous, irregularly 3 lobed
R Purplish
F Mauve, purple, pink or white. April to July
H A garden escape
   Perennial
   Poisonous

A most attractive plant which seeds itself liberally and which will grow almost anywhere, very quickly, flourishing in unwanted corners and the cracks in crazy-paving. The leaves are graceful and fern-like in spring (before the flowers) and their fresh green-blue colour and attractive form help to clothe the bareness of the early spring garden.

Various parts of the plant have been used in the past to cure diarrhoea, to promote perspiration, to help in childbirth, and to alleviate rheumatic pains; but this plant is a member of the all-poisonous Ranunculus family, and it should not be experimented with. However, dried crushed seeds, made into a dusting powder, will kill lice very effectively.

CONTROL Too pretty a plant to cast out – dig it up in winter or in the very early part of the year and plant in barren, rocky corners, under shrubs and trees and at the bases of old walls, etc.

CRUCIFERAE Cabbage Family

Honesty, Penny-flower
Lunaria annua L.

A Stiffly erect
AH 30–100 cm (12–40 in)
L Irregularly and coarsely toothed, acuminate, hairy
R Tap-rooted
F 4 petalled, magenta. April to June
H A garden escape
   Biennial

The pearly moons of Honesty's dried seed-pods are the most familiar of the 'winter everlastings'. There is an old saying that whoever grows this plant successfully is

exceptionally honest. It does not transplant happily after the seedling stage, so if it is wanted for winter flower decoration, sow seeds where the plants are to grow and thin out later. It will flourish almost anywhere, in sun or shade and looks well in the spring in association with forget-me-nots and narcissus.

CONTROL Easily weeded out by hand.

Aquilegia

136

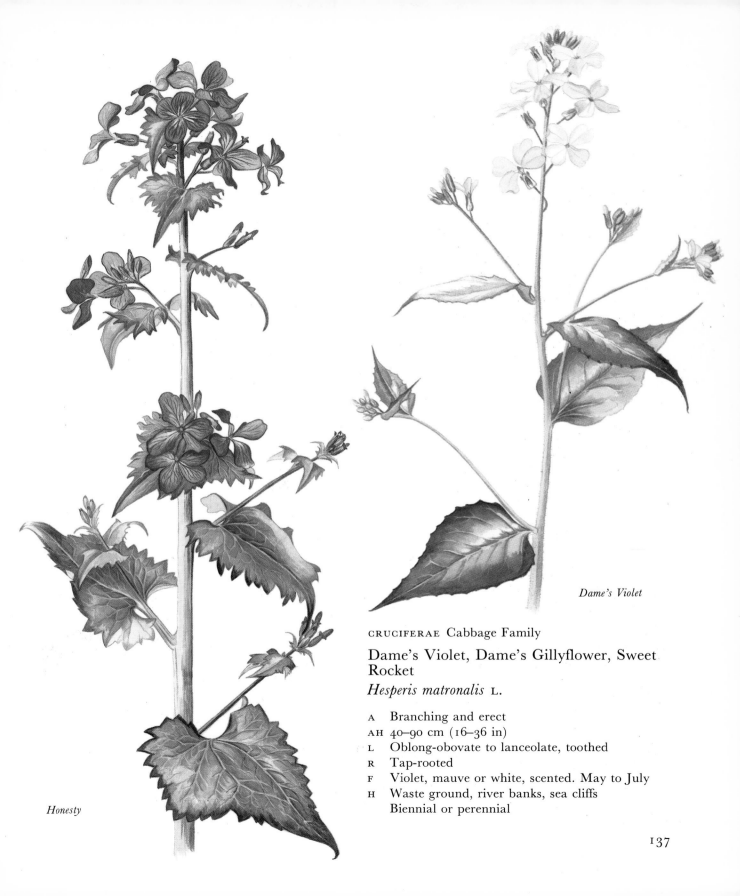

Dame's Violet

Honesty

CRUCIFERAE Cabbage Family

Dame's Violet, Dame's Gillyflower, Sweet Rocket
Hesperis matronalis L.

A   Branching and erect
AH 40–90 cm (16–36 in)
L   Oblong-obovate to lanceolate, toothed
R   Tap-rooted
F   Violet, mauve or white, scented. May to July
H   Waste ground, river banks, sea cliffs
     Biennial or perennial

137

A large, untidy, floppy plant, but worth transplanting if it turns up in an inconvenient part of the garden because of its sweet evening scent. Young plants may be moved in early spring, or it may be propagated from cuttings.

USES For a pot-pourri which includes Dame's Violet and which will retain its scent: gather about 2 lb (905 g) red Rose petals (these must not be gone-over blooms), cover with 8 oz (225 g) ordinary salt, not refined, and leave for a week. Add 8 oz (225 g) bay salt, 8 oz (225 g) powdered cloves and 8 oz (225 g) demarara sugar, 4 oz (115 g) gum benzoin, 2 oz (55 g) powdered orris-root, a large cupful of brandy and the dried petals of Dame's Violet, Lavender, Rosemary, Thyme, Bergamot, Lad's Love, and any other favourite flower.

CONTROL Easily removed by hand.

VIOLACEAE Violet Family

Heartsease, Wild Pansy, Monkey-faces, Three-faces-under-a-hood, Stepmothers, Love-in-idleness, Trinities
Viola tricolor L.

A   Much branched, sometimes trailing
AH 7–30 cm (3–12 in)
L   Glabrous, ovate, crenate, sometimes cuneate at base
F   5 petalled, blue, mauve, violet and yellow. April to September
H   Cornfields, cultivated land, waste ground
    Annual, sometimes perennial

Heartsease is adept at protecting itself from bad weather by drooping its head at night or at the onset of rain, so that its petals stay dry. It was formerly used as a remedy for epilepsy and diseases of the heart, and this probably accounts for its medieval name. It was Christianized in the sixteenth century, to become Trinities or Three-faces-under-a-hood.

This is the flower that Puck gave Oberon to work the mischief of 'A Midsummer Night's Dream', and it was often a vital ingredient in love-charms. The dried plant was powdered and put into the food or hidden in the room of the person desired.

It is the food plant of the gorgeous butterfly Queen of Spain Fritillary, *Issoria lathonia* L.

USES Heartsease is prescribed for diseases of the skin, such as acne, pruritus in the aged, spots, rashes, impetigo, eczema and psoriasis. The plant must be gathered when it comes into flower and quickly dried in the shade. Make a decoction as follows: to 2 pt (1.14 l) cold water add 2 oz (55 g) dried plant, and soak for 2 hours. Bring slowly to the boil, remove from heat and leave to infuse for 10 minutes. For all skin afflictions, drink at least 3 cups per day, between meals, and well before eating. At the same time apply compresses soaked in the same liquid to the affected areas.

CONTROL Heartsease is too pretty to destroy. Dig it up and put it where it may bloom undisturbed (it prefers a semi-shaded position).

PAPILIONACEAE Pea Family

Tufted Vetch, Blue Grass, Wild Vetch, Cow Vetch, Bird Vetch, Cat Peas
Vicia cracca L.

A   Scrambling
AH 60–200 cm (24–78 in)
L   Oblong-lanceolate to linear-lanceolate, tendrils branched
F   Blue-purple. June to August
H   Grassy and bushy places
    Perennial

This flower is generally seen scrambling along grassy roadside banks and over hedges. It appears a deep, brilliant blue when seen from a distance, but on closer examination, the fertilized flowers are seen to be magenta.

CONTROL Its twining, tendrilled leaf stems are easily unhooked and the plant removed to the compost-heap, but it is worth remembering that the nitrogen nodules on the roots will enrich the soil.

BUDDLEJACEAE Buddleia Family

Butterfly Bush, Buddleia
Buddleja davidii Franch.

A   Fountain-shaped shrub
AH 1–5 m (3–16 ft)
L   Ovate-lanceolate, acuminate, serrate, dark green above, white beneath

F Dense many-flowered cymes, lilac, mauve or purple. June to October

H Waste places and walls

This common though very attractive shrub looks its best when grown in isolation on a lawn, where the true fountain-shape of the bush may be properly appreciated. The graceful shape of the arching branches, tipped with spikes of purple in late summer, looks very different from the shrub often found growing squashed against a wall with all its main limbs chopped off in a vain attempt to keep it within bounds. The seedling trees will pop up between stones in paths, in cracks in steps, perch in gutters or even grow directly out of old walls, and the shrub became despised simply because of its ubiquitousness. Red Admiral butterflies are often seen covering the bush, though the caterpillars do not eat the leaves.

CONTROL Pull all the seedling plants out as soon as recognized before the tenacious roots damage walls, paths or paving.

APONCYNACEAE Periwinkle Family

Lesser Periwinkle, Myrtle
Vinca minor L.

A Procumbent and trailing

AH Low

L Lanceolate-elliptic, glabrous, dark green

R Stem rooting

F Blue-purple or mauve. March to May

H Woods, hedgebanks, roadsides
  Perennial

This is a herb of magic and love, one of whose earlier names was Sorcerer's Violet. In medieval times it was regarded as a love charm, when mixed with earthworms and Houseleeks, as described in *Secretes of Albartus Magnus of the Vertues of Herbs, Stones and certaine Beastes*: 'Perwynke when it is beate unto pouder with worms of ye earth wrapped about it and with an herbe called houselyke, it induceth love between man and wyfe if it bee used in their meales.'

And Apuleius in his *Herbarium* (1480) said: 'This wort is of good advantage for many purposes, that is to say, first against devil sickness and demoniacal possession and against snakes and wild beasts and against poisons and for various wishes and for envy and for terror and that thou mayst have grace, and if thou hast

the wort with thee thou shalt be prosperous and ever acceptable.'

Garlands of periwinkle were hung about the necks of those condemned to death in medieval Italy and England. The Italians called it 'Fiore di Morte' and placed it on the biers of their dead children as a symbol of immortality, perhaps because the plant is an evergreen.

This plant makes an excellent evergreen ground cover under deciduous trees, as it will suffocate all other weeds with its net of rooting, leafy stems.

USES Lesser Periwinkle has been used for hundreds of years to arrest haemorrhages. A decoction of 5 oz (140 g) plant to 2 pt (1.14 l) water (leave to soak for an hour, bring slowly to the boil, boil for $\frac{1}{2}$ minute and then allow to infuse for 15 minutes; dose: 3 cupfuls per day) is used today by homoeopaths in cases of malaria, leucorrhoea, anaemia, and excessive menstrual discharge. The plant is also used as a country remedy for pleurisy in the form of equal quantities of expressed juice and white wine in a wineglass.

As a remedy for mouth ulcers, tonsillitis and as a gargle, a decoction (2 oz (55 g) plant to 2 pt (1.14 l) water) may be made. A fresh leaf of the plant squashed up and pushed into the nostril is a cure for nosebleeding.

CONTROL The trailing stems root about every 10 in (25 cm) and the area will have to be thoroughly dug and sifted through for all the broken pieces of root.

BORAGINACEAE Borage Family

Comfrey, Knit-bone, Bone-set, Abraham Isaac and Jacob
Symphytum officinale L.

A Stout and erect

AH 30–120 cm (12–48 in)

L Leaves ovate-lanceolate, winged, hispid, veined

R Thick, fleshy, fusiform

F Purple, blue, mauve, crimson, pink or whitish. June to July

H Damp places
  Perennial

A book could be written about this plant which has been used medicinally for over 2,000 years. It is one of the few members of the vegetable kingdom to contain allantoin (which stimulates healthy granulation of

Heartsease

Tufted Vetch

Butterfly Bush

Lesser Periwinkle

Comfrey

tissues), as well as minerals such as manganese, iron, calcium and phosphorus, proteins, the element iodine and the vitamins A, B, C and B$_{12}$.

In the Middle Ages Gerard said of it: 'The slimie substance of the roote made into a posset of ale, and given to drink agaynst the paine in the backe, gotten by any violent motion as wrestling, or over much use of women, doth in fower or five daies perfectly cure the same, although the involuntarie flowing of the seed in men be gotten thereby.'

USES In Bavaria, Comfrey leaves are made into fritters (that look like green fish): wash some fresh young Comfrey leaves, cut off as much stem as possible, dip in pancake batter and fry.

The ailments for which Comfrey is traditionally a cure are legion, and a short list includes diarrhoea, dysentery, enteritis, stomach ulcers, gout, phlebitis, burns, scalds, bronchitis, laryngitis, catarrh, smoker's cough, erysipelas, external ulcers, slow-to-heal cuts, sores and wounds but, finally, its best-known use is to aid broken bones to knit cleanly. (The name *Symphytum* is derived from the Greek word meaning 'to unite'.)

As a remedy for diarrhoea and the allied troubles of enteritis and dysentery, prepare as follows: soak 4 oz (115 g) minced or chopped root of Comfrey in 4 cups of cold water for 3 hours: strain. Dose: this quantity per day. This same preparation is effective in cases of stomach ulcers, as it helps the growth of new tissues in the stomach lining.

As a compress for all external uses, such as sprained and swollen wrists, or healing wounds, burns and scalds, painful joints, phlebitis, varicose veins and insect bites, the remedy is fresh ground or grated root, mixed to a paste with water, applied on a clean dressing kept in place with a bandage.

CONTROL Dig up the large and lumpy roots, taking care to capture all the broken pieces. It makes superb compost.

SOLANACEAE Nightshade Family

Deadly Nightshade, Dwale
Atropa bella-donna L.

A   Erect and bushy
AH   Up to 150 cm (60 in)
L   Ovate acuminate, glabrous
F   Brownish-violet or yellowish, bell-shaped. June
     to August

Deadly Nightshade

H   On calcareous soils in woods and hedges
    Perennial
    Poisonous

This is the most well-known, but least seen, member of the Nightshade family, since the plant is now becoming scarce. Legend and myth surround it, and Gerard advised: 'If you will follow my counsell, banish it from your gardens and the use of it also, being a plant so furious and deadly.' Its name is explained by a Venetian physician who said that the beautiful ladies of Venice distilled a water from the leaves which they then dropped into their eyes to enlarge the pupils and make them more lustrous and beautiful – hence 'bella donna' (beautiful woman). Deadly Nightshade was known as Dwale in Chaucerian times, and there are several suggestions as to the derivation of the word – the French word 'deuil' or grief, and the Norse word 'dool' or delay, sleep. The generic name *Atropa* is taken from the Greek 'Atropus', who was one of the three Fates: she it was who held the shears that cut the thread of life.

The plant has been used for murder throughout history, and there is a tale of how the soldiers of Macbeth, serving under King Duncan the first of Scotland, arranged a truce with the invading Danes. However, Macbeth's men prepared a great quantity of liquor, laced with Dwale, to refresh the thirsty Danes, who drank deeply of the welcome brew and, somnolent and stupefied, were then slaughtered in heaps.

Bella-donna is grown commercially in southern Europe for medicinal purposes, and a liquid extract is made into liniment, plaster, ointment, suppositories and tinctures. It is a most valuable herb in conditions of the eye, and it is still used in eye operations. It must *never* be experimented with, as it is said that three ripe berries are sufficient to kill an adult. The root is the most poisonous part of the whole plant, containing atropine, hyoscyanine and scopolamine (hyascine), then the leaves and, lastly, the famous black berries.

CONTROL Once you have recognized this striking, rather sinister-looking plant, think carefully before you dig it up and burn it, because it is beautiful and becoming rare.

SOLANACEAE Nightshade Family

Woody Nightshade, Bittersweet
Solanum dulcamara L.

A   Scrambling
AH Up to 200 cm (78 in)
L   Hastate
F   Purple with cone of golden yellow anthers. June to September
H   Hedges, waste ground, shingle
    Perennial
    Poisonous

This very common plant is often, mistakenly, called Deadly Nightshade and though its red berries are very poisonous, they are not as deadly as the shining black berries of *Atropa bella-donna* (the true Deadly Nightshade). Woody Nightshade is sometimes called Bittersweet because the berries taste intensely bitter at first, but rapidly change to a sickly sweetness. The plant was used by herdsmen as an amulet against magic, and twisted garlands of it were hung about the necks of pigs, sheep and cattle to protect them from witchcraft. In the tombs of Tutankhamen, dried berries of Bittersweet were discovered worked into a jewelled collar which was part of the grave furniture.

USES Though it can be dangerous, the plant has been used by skilled herbalists for thousands of years. A safe, present-day use for it is to apply the juice extracted from the crushed leaves to facial blemishes.

CONTROL Do not fear too greatly that young children will sample the berries. They are poisonous if taken in great quantity, and the taste alone should prevent this. However, if the presence of the plant worries you, the perennial rootstock must be grubbed up, which may prove difficult if the root is entwined in hedge-stems. If so, keep cutting off all the young shoots as they appear, until the plant eventually dies.

SCROPHULARIACEAE Figwort Family

Purple Toadflax
Linaria purpurea (L.) Mill.

A   Tall and branching
AH 30–90 cm (12–36 in)
L   Linear or linear-lanceolate, glabrous, glaucous
F   Dense racemes of purple flowers. June to August
H   Walls and waste places
    Perennial

Woody Nightshade

Purple Toadflax

145

This tall, fine-leaved plant may suddenly appear in the border uninvited and once you allow it to seed, the purple spikes will be with you for ever. However, it is an attractive plant, which will grow in hot, dry conditions where sturdier border plants will fail, and if you dig up the seedlings once recognized and plant them all together, they will provide interest in the border in late summer.

CONTROL Easily weeded out by hand.

SCROPHULARIACEAE Figwort Family

Ivy-leaved Toadflax, Monkey-mouths, Mother-of-millions

Cymbalaria muralis Gaertn., Mey. & Scherb. (*Linaria cymbalaria* (L.) Mill.)

A   Trailing and drooping
AH 10–80 cm (4–32 in)
L   3–7 lobed, thick, sometimes purplish beneath
F   Mauve with a yellow spot. May to September
H   Walls
    Perennial

Ivy-leaved Toadflax

This delightful and delicate little plant was carefully cultivated as an exotic in the early seventeenth century, then escaped to colonize old walls everywhere. It has a most effective and methodical way of propagating itself; when the seed-capsule is ripe, its pedicel bends over the wall's surface, seeking a crevice. When it finds one, the capsule bursts and the seeds are discharged into the comfortable bed of old mortar in which they will germinate. The unfertilized flowers display themselves prominently above the leaves, while the fertilized blooms retreat to the shelter between wall and leaf-canopy to develop the capsule. The early herbals of Parkinson and Gerard show plates with the *Linaria cymbalaria* springing upwards from a brick wall, probably because the blockmakers were unwilling to accept that this plant grows downwards.

CONTROL The stems are surprisingly wiry, but the plant can easily be pulled out of its crevices in a wall, which will need re-pointing to prevent the seeds germinating.

LABIATAE Labiate Family

Mint, Spearmint
Mentha spicata L.

A Tall and vigorous
AH Up to 1 m (3 ft)
L Oblong-lanceolate, serrate, glabrous, very aromatic
F Cylindrical spike of mauve flowers. August to September
H A garden escape, roadsides and waste places Perennial

The innumerable varieties of mint have been used throughout history for flavouring both savoury and sweet dishes and as a vital ingredient in wine. The plant was even used as a perfume, particularly for the arms, in ancient Greece, where at one time an edict was issued to forbid soldiers, during times of war, to include mint in their rations as 'it did so much to incite Venery, that it tooke away or at least abated their animosity or courage to fight'. (Parkinson)

The generic name is derived from the nymph Mentha, who was beloved by Pluto but who was changed into a plant by the bitterly jealous Proserpine.

A small clump of Mint is very useful even though it tends to smother everything else in its vicinity. If you have no Mint at all, buy a small bunch, put it in a glass jar of water in a good light, and leave it for about 3 weeks, topping up the water from time to time. Small white rootlets will appear and the plant can be transferred to a pot of damp soil, which should be watered well for the first month. Plants affected by mint-rust should be destroyed, or they may affect the whole patch.

USES The plant is particularly useful in the kitchen. (Put some sprigs of Mint on the shelves of the larder, particularly near the window, to repel ants in the summer.) It is delicious added to the first new potatoes of the year but there are many other uses for the herb, such as Mint jelly: steep a handful of Mint leaves in $\frac{1}{2}$ pt (0.3 l) hot, made-up gelatin, and strain before allowing to set. The jelly should be a pleasing shade of green.

Mint tea, which is part of the daily ritual from North Africa to Turkey is prepared (with local variations) as follows: take approximately 8 tbsp chopped or torn-up Mint leaves to 1 pt (0.6 l) boiling water, put the Mint into a crockery (not metal) teapot and pour boiling water over it, as in making tea. Allow to infuse, strain well and pour into glasses, garnished with a sprig of fresh Mint and/or a slice of lemon. Sweeten (or not) to taste.

CONTROL Mint must be controlled. Plant it in a concrete-surrounded area, or in an old enamel pail or tin bath with the bottom knocked out, to prevent it monopolizing your herb bed. It can be eradicated by mulching with bonfire ash, which it hates but which will not hurt anything else. Otherwise, ruthlessly pull out the rooting runners in winter.

Self-heal

Mint

LABIATAE Labiate Family

Self-heal, All-heal, Blue Curls
Prunella vulgaris L.

A   Erect
AH 5–30 cm (2–12 in)
L   Pubescent, ovate, entire or shallowly dentate,
    cuneate
F   Violet. June to September
H   Grassland, wood clearings, waste places
    Annual

The vivid purple flowers of Self-heal spring up in areas of rough grass in late summer and indicate the need for plenty of natural fertilizer to improve the soil. The name is a corruption of the medieval German word *Brunella* derived from *Die Breuen*, an affliction of the mouth, for which this plant was used as a remedy.

USES As a gargle for throat infections, such as thrush, stomatitis, pharyngitis and mouth ulcers, boil 1 oz (30 g) herb in 1 pt (0.6 l) water, strain, cool and sweeten to taste, flavouring with lemon if liked. It is quite safe to swallow, and even tastes nice.

CONTROL It is easily hand-weeded.

148

LABIATAE Labiate Family

Ground Ivy, Alehoof, Gill-go-by-the-ground
Glechoma hederacea L.

A Creeping
AH 10–30 cm (4–12 in)
L Reniform to ovate cordate, obtuse, crenate,
    softly hairy
F Violet with purple spots. March to May
H Woods, grassland and waste places
    Perennial

No relation of the wall-clothing Ivy (*Hedera helix*), this plant was called Alehoof because it was added to beer to clarify, preserve and flavour it – the old English word *hofe* is a brewing term. Gill-go-by-the-ground is from the French word *guiller* – to ferment. The well-named

Ground Ivy can cover the ground to such an extent that it prevents anything else growing, and it is sometimes seen as a complete carpet in oak woods, especially where coppiced.

USES Gill tea may be made as a cough remedy with 1 oz (30 g) leaves (usually to be found all year round) to 1 pt (0.6 l) water. Sweeten, and if liked, flavour with liquorice.

The juice of the plant, sniffed up the nose, will help to cure headaches and migraine, and is also good – applied externally – for bruises and black eyes.

CONTROL The plant is very easily weeded out by hand, but all the trailing, rooting ends must be gathered up.

DIPSACACEAE Teasel Family

Teasel, Barber's Brush, Venus' Basin
Dipsacus fullonum L.

A Tall, stout and branched
AH 50–200 cm (20–78 in)
L Oblong to elliptical-oblanceolate, glabrous,
    prickly, connate, crenate
R Yellowish tap-root
F Rose-purple, in bands. July to August
H River banks, rough pasture
    Biennial

This is not the plant whose spiny dried heads are still used to comb cloth – the variety which is cultivated for this purpose is *D. sativus*. Teasel collects rain in the large leaves which clasp the stem, and according to country superstition the resulting fluid is a remedy for warts and is good for bathing tired eyes and beautifying the complexion. The name 'Venus' Basin' was given to the plant because the fluid was supposed to make maidens as beautiful as Venus herself. If any hungry insect tries to climb past the barrier of this herbal moat, it will drown and the bottom of this natural cup usually harbours insect parts which have not dissolved.

In the flower-heads of the Teasel lives a small grub, which in Elizabethan times was worn about the neck as

Ground Ivy

Teasel

a charm against the quartan ague, a fever characterized by the occurrence of a paroxysm every fourth day.

The dried seed-heads make long-lasting flower arrangements.

CONTROL The young leaf-rosettes resemble a large, clear-green, prickly Cos-lettuce. Root them out or the resulting plant will smother its near neighbours. Alternatively, move them when very small (they have a tap-root and dislike disturbance) into a clump, where the rigid stems and flower-heads can be seen to advantage, particularly throughout the winter months.

COMPOSITAE Daisy Family

Burdock, Hurr-burr, Beggar's Buttons
Arctium lappa L.

A   Large-leaved clumps
AH 1–2 m (3–6 ft)
L   Large, folded, ovate-cordate, green above, grey-cottony beneath
F   In subcorymbose clusters, hemispherical. July to September.
H   Waste places
    Biennial

This is a huge plant when fully grown, with enormous felted and folded dull-green leaves, topped with hooked seed-heads which get caught up in clothes and animal fur and are so dispersed. The artist Stubbs often

Burdock

painted this familiar plant into the background of his equine masterpieces.

The Burdock has a reputation as a love-potion, having 'grete strength in veneriall pastymes'. The recipe called for the seed-heads to be pounded and mixed with the private parts of a goat and some hairs from a white puppy (cut on the first day of the new moon and burnt on the seventh day). This was mixed with brandy and left uncovered for exactly 3 weeks in order that the mixture should become impregnated with the astral influences, then cooked until thick, with a few drops of crocodile sperm. There may well be some truth in the legend since Gerard said that the stalks, peeled and eaten raw 'do increase seed and stir up lust' and one of the plant's older names was 'Love Leaves'.

USES An infusion may be made of 2 oz (55 g) cut fresh root to 2 pt (1.14 l) boiling water. Allow to cool, add honey and 1 pt (0.6 l) milk. Dose: drink half first thing in the morning and the rest last thing at night. This is reputed to be excellent for rheumatism, gout and arthritis, particularly if at the same time poultices of the fresh leaves, boiled for a very short time in as little salted water as possible, are applied to the affected areas. This poultice can also be used to alleviate bruises and swellings.

For thinning hair, make a lotion of 1 oz (30 g) fresh leaves to 2 pt (1.14 l) water and massage the head with it daily.

CONTROL Pull up the plant as soon as recognized. Put only the leaves into the compost-heap; the woody stems will take too long to rot down.

COMPOSITAE Daisy Family

Spear Thistle
Cirsium vulgare Savi. Ten.

A   Stiff and spiny
AH 30–150 cm (12–60 in)
L   Deeply pinnatifid, undulate, lobed, toothed and
    tipped with stout spines. All prickly-hairy
R   Long tap-root
F   Ovoid-oblong, purple-red. July to October
H   Fields, waysides, waste places
    Biennial

The Scots supposedly chose the thistle for their national emblem after an unwary raiding Dane stepped on one and howled in agony, thus raising the

Spear Thistle

Creeping Thistle

152

alarm. The Scots were then able to defeat the marauders, and the 'guardian thistle' was given the suitable motto *Nemo me impune lacessit*, 'No one provokes me with impunity'.

CONTROL Despite the rhapsodies of Robert Burns:

> The rough burn-thistle spreading wide
> Amang the bearded ear—
> I turned the weeding heuk aside
> An' spared the emblem dear!

Thistles should be dug up as soon as the leaf-rosettes are recognized, before the spines of the adult plant can penetrate your gardening gloves.

COMPOSITAE Daisy Family

Creeping Thistle
Cirsium arvense (L.) Scop.

A   Stiff and spiny
AH 30–120 cm (12–48 in)
L   Glabrous, cottony, pinnatifid, undulate, triangular-toothed with spiny ciliate lobes
R   Tap-root with far-creeping lateral roots
F   Dull pale purple. July to September
H   Fields, waysides, waste places
    Perennial

This variety of thistle is not even architecturally interesting, as are some other members of the Thistle family, and it is best extirpated as soon as recognized. It is interesting botanically in that the male and female plants are separate; the Creeping Thistles in one patch will be male and in the next female, but both groups are connected by the same underground stem, which penetrates very deeply into the soil.

It is the food plant of the caterpillars of the beautiful Painted Lady butterfly, *Vanessa cardui* L.

CONTROL If the thistles are growing in a patch of rough grass, the only thorough method of eradication is to rotovate and re-seed – chopping off at ground-level only seems to encourage the thistles to proliferate with greater vigour, much like pruning roses.

IRIDACEAE Iris Family

Stinking Iris, Gladdon, Gladwyn, Roast-beef Plant
Iris foetidissima L.

A   Clumps of spear-shaped leaves
AH 30–80 cm (12–32 in)
L   Evergreen, ensiform
R   Rhizomatous
F   Purplish-livid. May to July
H   Hedgebanks, open woods and sea cliffs
    Perennial
    Poisonous

Stinking Iris

The leaden-coloured flowers of this plant are almost invisible and have a very short flowering season. The berries, however, are bright and long-lasting and *Iris foetidissima* is worth cultivating just for these. Legend has it that you should draw a circle round this plant three times with a double-edged sword to keep away the earth-demons to whom this plant belongs and if you take it from the ground, a payment of wheat-cakes should be left in its place.

The name Stinking Iris is derived from the fact that it smells distinctly (of roast beef – and not unpleasant) when handled. The curious name of Gladdon is from the Old English word *Glaedene* which, in turn, was taken from the Latin *Gladiolus*, a little sword, which the leaf-shape resembles.

This iris grows easily from seed; add some chalky soil to the compost, and more when you plant out the seedlings. They prefer semi-shade beneath deciduous trees, but will tolerate full sun and arid conditions though not so many flowers will be produced.

CONTROL When the plants are mature they are quite difficult to dig up, but it is worth transplanting to a less-favoured spot because of the brilliance of the winter berries. Move it in winter.

Green-flowered species

Horsetail

EQUISETACEAE Horsetail Family

Horsetail, Cat's Tail, Bottlebrush, Shavegrass, Pinegrass

Equisetum arvense L.

A   Like a small Christmas tree
AH 2–8 cm (1–30 in)
L   None
R   Rhizomatous
F   None
H   Waste places
    Perennial

This miniature relic of time before man is a survivor in every sense of the word and will pop up, blithe and joyous, in the most inhospitable places. Horsetail is so scaly in texture that it was used in earlier days as a pot-scourer and even by cabinet-makers as a natural sander–polisher.

USES A totally natural foliar-spray can be made to feed the roses and combat mildew: take two large handfuls of the plant, put in a large saucepan and cover with cold water. Bring to the boil slowly, turn heat down and simmer gently for 20 minutes. Cover and set aside to cool until the next day, then strain. Use this as the concentrate and dilute 2 parts water to 1 part concentrate. This will help to combat Hollyhock and mint-rust.

A poultice of the crushed plant will help to heal cuts and wounds more quickly, and the juice from washed, crushed plants may be taken for stomach haemorrhage.

CONTROL Exceedingly valuable in the compost-heap if this is hot enough to kill the roots, as Horsetail has a high content of silica, cobalt and calcium. If your compost-heap is not hot, burn the Horsetail and add the ashes to the heap. Horsetail likes poor, dusty ground so if your garden is infested with it, enrich, mulch, fertilize, manure, and add peat, and leaf mould if you can get it, to your soil.

POLYPODIACEAE Fern Family

Bracken, Brake Fern, Eagle Fern, Upland Fern

Pteridium aquilinum (L.) Newm.

A   Fern-like
AH 15–400 cm (6–160 in)

Bracken

L 3 pinnate, deltoid, pinnae lanceolate or oblong,
  segments pectinately arranged
R Rhizomatous, far-creeping
F None
H Woods, heaths
  Perennial
  Poisonous

Bracken is among the first of the plants to repopulate
burnt woodland or heath and in the warm, damp,
growing time of spring, the 'shepherd's crooks' of the
Bracken may actually be seen pushing their way
upwards, so quickly do they grow initially. A most
important plant among the witches' collection of magic
herbs, it was said that if you gathered the tiny, rusty-
dark spores on St John's Eve you could become invisible at will and discover the secret of eternal youth.

It is the food plant of the Broom moth, *Ceramica pisi* L.

USES Bracken contains a great deal of potash, particularly in early summer, when piles of green Bracken
can be collected and burnt, and the ash spread on
gardens with light loam and sandy, gravelly soil. It
makes good litter to protect tender plants in winter,
and can be used instead of straw, to line strawberry-
beds where it lasts longer and the scratchy edges deter
slugs and snails.

CONTROL Where lawns have been laid over an old
brake (without first grubbing out the roots), constant
mowing serves only to control the growth. Bracken
roots can be 60 cm (2 ft) down, in solid blocks of
rhizomes and this daunting prospect can be properly
dealt with only by deep rotovating, or by use of chemicals, with a careful watch afterwards for the re-
emergence of new fronds. Keep cutting these off, and
you stand a fair chance of winning! No frost will ever
kill off the root system.

POLYPODIACEAE Fern Family

Common Fern, Broad Buckler Fern
Dryopteris dilatata

A Fern-like
AH 30–170 cm (12–68 in)
L Dark green, triangular-ovate to lanceolate,
  segments oblong, toothed, pinnately lobed
R Rhizome
F None

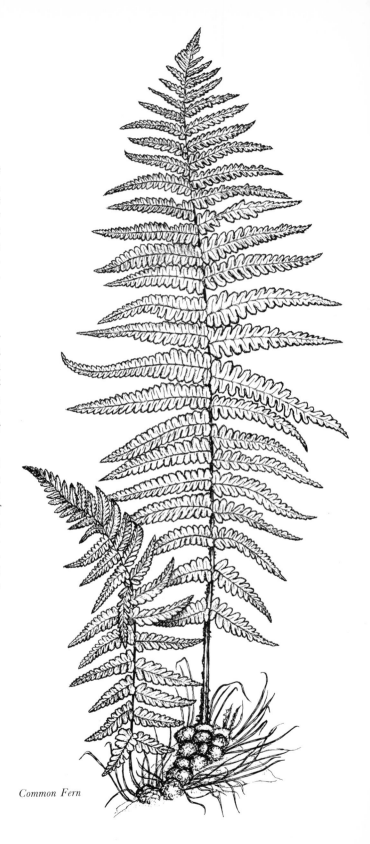

Common Fern

Many-seeded Goosefoot

H   Woods, hedges, shady rock ledges
      Perennial

This is included here as one example of the many ferns that may lodge themselves in the damp crevices of walls and steps. The graceful fronds of ferns were very popular in nineteenth-century conservatories, and the hardy species seem to be coming back into fashion. However, it is impossible to move an adult plant to a more appropriate place if, as is usual, it is growing at the bottom of a drainpipe.

CONTROL Easily sheared off.

CHENOPODIACEAE Goosefoot Family

Many-seeded Goosefoot, All-seed
Chenopodium polyspermum L.

A   Erect
AH   Up to 1 m (3 ft)
L   Ovate or elliptic, with a single tooth-like angle
      on one or both sides
F   Lax cymes. July to October
H   Waste places, cultivated ground
      Annual

If part of the garden has been rotovated, All-seed, in great patches, may well be one of the first visitors to try out the new hotel. It is an undistinguished plant but has been valuable to man as food for thousands of years. Seeds of it formed part of Grauballe man's last meal before he was ritually murdered and buried in the peat bogs of Jutland in Denmark around 2,000 years ago; it is supposed that this sacrifice was part of an annual rite to ensure the fertility of the land in the year to follow.

CONTROL Easily pulled out of the ground, and valuable in the compost-heap. Unlikely to occur in town gardens.

CHENOPODIACEAE Goosefoot Family

Fat Hen, All Good
Chenopodium album L.

A   Erect and mealy
AH   Up to 60 cm (24 in)
L   Rhomboid to ovate-lanceolate, toothed, deep
      green

159

F   Dense green cymes. July to October
H   Waste places, cultivated land
    Annual

Fat Hen is happiest growing in the muck of farmyards, and was still used early in this century as a food plant in Scotland and Ireland.

USES Fat Hen can be cooked like spinach, adding pepper, salt, and a little butter or bacon fat before serving. The plant contains more protein and iron than spinach.

CONTROL Easily weeded out by hand.

ACERACEAE Maple Family

Sycamore, Lock-and-key-tree
Acer pseudoplatanus L.

A   Large tree with broad spreading crown
AK Up to 30 m (100 ft)
L   5 lobed, cordate, dark green and glabrous
    above. Lobes coarsely and irregularly
    crenate-serrate
F   60–100 in a narrow pendulous panicle. April to
    June
H   Woods, hedges, cliffs

The Sycamore is well known for its winged seeds which can be seen spinning away from the parent tree in autumn gales. Each 'propeller' will result in a quick-growing, lusty young tree, which can start life anywhere, even on or out of walls and neglected guttering. However, the Sycamore prefers deep, rich, moist, well-drained soil, and the new leaves of a young tree in ideal conditions are a shining purple-bronze which turn to a bright, dark green.

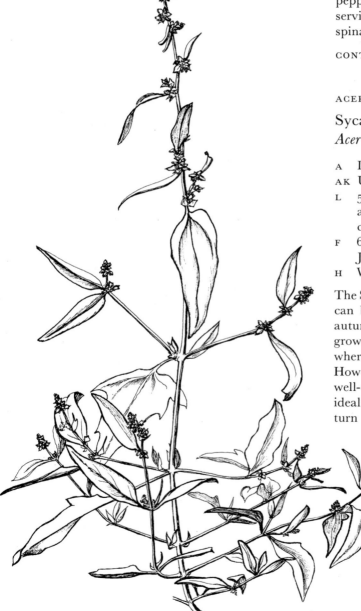

Fat Hen

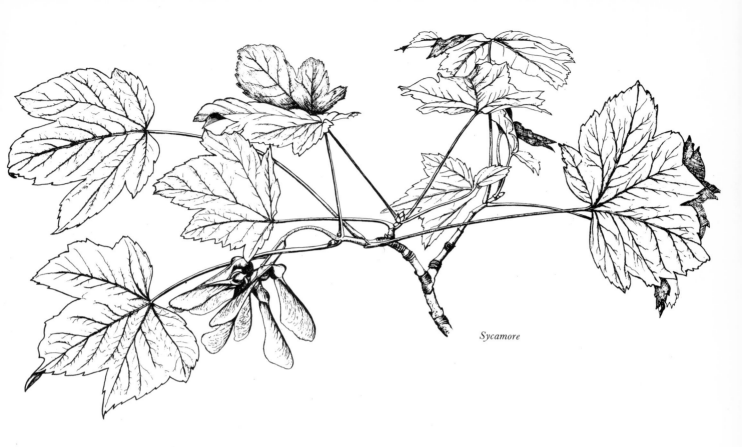

Sycamore

Trees as vigorous as the Sycamore are to be discouraged near houses, walls and drives, as their roots will disturb foundations, mains and drains, and crack and lift tarmac. As a general guide, the area of a tree's roots below ground is approximately the same as its branch-area above.

The bark of the tree was once used to make a lotion for tired eyes and skin irritations, and the sappy 'inner bark' is clean and wholesome and may, in emergency, be used as a dressing for wounds.

CONTROL The young seedling trees should be lifted out of the ground or wall before the roots get a hold in the subsoil. If a large tree has been cut down, a forest of suckers will spring up around the base of the old trunk. The stump itself should be treated with a chemical stump-killer if there is no room to drag the whole stump out, roots and all.

ROSACEAE Rose Family

Parsley Piert, Field Lady's Mantle, Parsley Breakstone, Parsley Piercestone

Aphanes arvensis L.

A   Inconspicuous and low
AH 2–20 cm (1–8 in)
L   Pilose, fan-shaped, trisect
F   Very small in sessile clusters. April to October
L   Arable land, grassland
    Annual

Parsley Piert is no relation of the true Parsley. The first part of its name refers to the cut-up appearance of the leaves, and the latter half to its ability to break through hard and stony ground, derived from the French 'perce-pierre'.

USES This tiny plant has been used for many hundreds of years by herbalists as a cure for bladder stones, and as a general remedy for jaundice, dropsy, gravel and similar complaints of the bladder and kidneys. It is a strong but safe diuretic and may be made into an

Parsley Piert

infusion – a handful of the herb, fresh if possible, to 1 pt (0.6 l) boiling water. Drink this quantity daily in 4-oz (115-g) doses.

According to Gerard's herbal of 1597 the herb also 'cures Gonnorrhoea and ulcers and inflammations of the privie or secret parts of man or woman'.

CONTROL Leave it since so small a plant can hardly be seen, and will be minute in lawns made on dry soils.

ARALIACEAE Ivy Family

Ivy

Hedera helix L.

A   Climber
AH Up to 30 m (100 ft)
L   Glabrous, dark green, palmately 3–5 lobed
R   Stem-rooting
F   Subglobose umbels, yellowish green. September to November
H   Woods, hedges, walls
    Perennial

Ivy is generally disliked because it suffocates its host, and is accused of pulling mortar out of walls. However, in the Ivy's favour, its flowers are almost the only source of nectar for bees in the winter and on fine days late in the year the flowering umbels are almost alive with insects of many kinds.

Ivy was supposed to prevent drunkenness, and old taverns had drinking goblets turned from the thick stems of the Ivy. Early pictures of Bacchus show him with a wreath of Ivy instead of the vine-leaves that he is subsequently portrayed with. A drink made from a few Ivy berries crushed and slowly boiled in wine was recommended as a cure for the medieval hangover, but this remedy is not suggested for our more delicate modern systems, as research on Ivy juice indicates that it might kill human red blood-corpuscles.

Ivy is the food plant of the Oak Eggar moth, *Lasiocampa quercus* L.

USES A stain remover for navy and black wool garments, such as uniforms, can be made from Ivy. Use an old iron or enamel saucepan, fill it with Ivy leaves, cover with cold water and bring to the boil; keep boiling for 30 minutes. Lower the heat and simmer for about 3 hours. Strain and measure the remaining liquid. Add 1 tbsp ammonia to 1 pt (0.6 l). Put the liquid into a labelled bottle, and use to remove marks from serge, gabardine, barathea and similar materials.

As a cure for corns, soak an Ivy leaf in lemon juice or vinegar for 3 hours. After a lengthy hot bath, or after soaking the feet in hot water, apply the prepared leaf and bandage it in place, changing the leaf daily for a similar one, until eventually the corn comes away.

Compresses and poultices, either hot or cold, may be applied with good effect for neuralgia, neuritis, phlebitis, rheumatism and sciatica. For a cold compress of fresh leaves, chop and mince fresh Ivy leaves and apply direct to the affected area, keeping in place with bandages. For a warm poultice, take 2 handfuls fresh leaves, mince them, and add to 4 oz (115 g) bran in ½ pt (0.3 l) water. Mix together and warm gently in a saucepan – do not boil. Spread the paste on gauze of lint and apply to the affected area. Keep in place until the poultice is quite cold.

CONTROL To dislodge a mature Ivy from a full-grown tree, sever the thick stem or stems near to the ground, and pull them well away from each other and the host. When the Ivy leaves wither and eventually discolour, they can be distinguished from the tree's own leaves and the task of stripping off the rest of the plant can begin.

If you do not want Ivy choking your apple trees, seek out the seedling plants and destroy them. However, remember that the evergreen, shining leaves of Ivy can be immensely useful and ornamental in concealing ugly garden sheds roofed with rusting corrugated iron, your too-new asbestos garage, the central-heating oil-tank or (with the aid of a wire fence) your next-door neighbour. Ivy grows very quickly indeed and, being an evergreen, it is better than Russian Vine (*Polygonum baldschuanicum*) for camouflaging purposes.

CUCURBITACEAE Gourd Family

White Bryony, Red Bryony, Death Warrant, Jack-in-the-hedge, Tetterbury, English Mandrake

Bryonia dioica Jacq.

A   Climber
L   Palmately 3–5 lobed, lobes sinuate-toothed
R   Tuberous and massive
F   Greenish flowers in axillary cymes. May to June

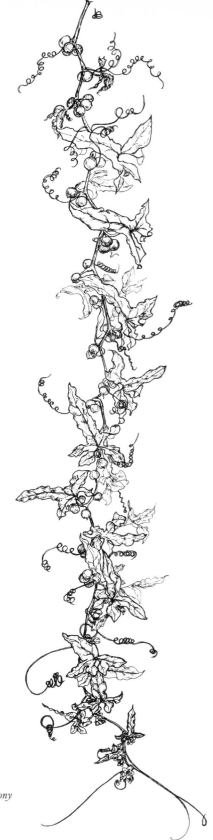

White Bryony

163

Ivy

H Hedgerows, scrub
Perennial
Poisonous

The roots of this plant are perennial but the tendrilled stems are annual, and grow to a great length. The root grows to an enormous size in maturity – 'of the bigness of a child of a yeere old' as Gerard described it.

The plant was called English Mandrake at one time, because the root could be roughly carved into a human figure, the shape of the true Mandrake root, and sold in herbalists' shops as the latter. Genuine Mandrake root (*Mandragora officinale*) was sought after because it was considered to help barren women conceive.

The root itself (when it was not masquerading as a Mandrake) was called 'Devil's Turnip' and in France fatal accidents have occurred because country people believed that eating the root would diminish milk secretion at the time of weaning their babes. The name Tetterberry originated in the Middle Ages because beggars were said to use the plant to cure their sores or 'tetters'. As is now known, the juice has exactly the reverse effect and it is likely that they deliberately made their sores look worse in order to excite pity, and alms.

This dangerous plant (an intake of fourteen berries has caused death in children, but it would need approximately forty to kill a healthy adult) was used medicinally by the Greeks, Romans and Egyptians. The juice of the plant is a powerful irritant to sensitive skins and will cause redness and blisters.

CONTROL Since the roots are impossible to dig out from the hedges where they are most usually found, the trailing, tendrilled growths should be constantly cut off, until they eventually cease to grow. However, if an old hedge laced with White Bryony should be bulldozed out, great care should be taken not to leave the poisonous roots exposed where cattle might eat them.

EUPHORBIACEAE Spurge Family

Dog's Mercury
Mercurialis perennis L.

A Erect and ground-covering
AH 15–40 cm (6–16 in)
L Elliptic-ovate, crenate-serrate
R Rhizomatous
F Green and catkin-like

Dog's Mercury

165

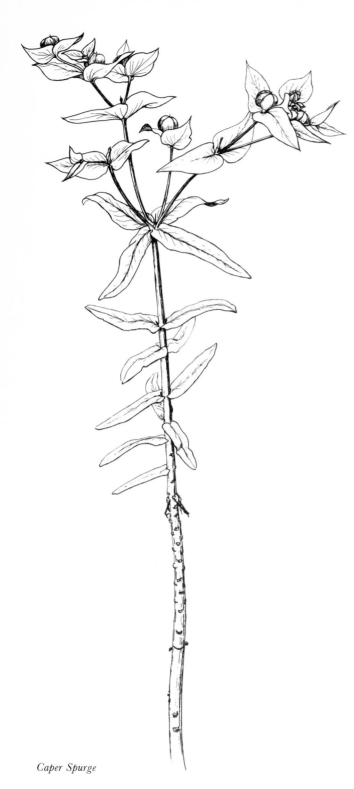

Caper Spurge

H Woods
Perennial
Poisonous

A light-green, leafy plant with dangling green strings of flowers, it is one of the first to clothe the bare woodland earth in spring. Legend has it that, since male and female plants are separate, the sex of a newly conceived child could be determined by taking the juice of the appropriate plant for three days after the presumed conception. It must be remembered that Dog's Mercury is poisonous and should on no account be experimented with – fatalities are unlikely, but very unpleasant symptoms of nausea, abdominal pains and diarrhoea will certainly result.

The first part of its name derives from the fact that plants which were generally undistinguished or lacking in scent, for example, Dog Violet, Dog Rose (very little scent), Dogwood and Dog Daisy, were all given the prefix of 'dog'.

USES As a wart cure, the fresh juice of the plant, collected when in flower, may be mixed with sugar or vinegar and applied to the affected area, taking care not to allow the compound to touch the skin around the wart.

CONTROL Dog's Mercury spreads more by its creeping roots than by the seed, so care should be taken to collect up as many of the roots as possible. It does not like being disturbed and will gradually die off with continuous hoeing.

EUPHORBIACEAE Spurge Family

Caper Spurge, Mole Plant
Euphorbia lathyrus L.

A Tall and rigid
AH 30–120 cm (12–48 in)
L Glabrous, glaucous, opposite-leaved, narrow-oblong, mucronulate, white-veined
R Tap-rooted
F Very small, 'flowers' are bracts. May to June
H Woods
Biennial
Poisonous

This strange, quadrangular plant, with its rigid stem and smooth, blue-green leaves standing out at right angles from the stalk, is thought to repel moles. The plant is also named for its fruits, which split open noisily on hot sunny days, flinging the seeds in all directions. These fruits have been pickled as 'capers' in the past and it is amazing that few accounts exist of the consequent effects of being poisoned, even though fatalities would not be likely. The plant was formerly grown in herb gardens as a medicine-plant, but is no longer used as its purgative action is too violent.

It makes a very handsome and architectural plant for the garden, with its exactly regular, white-veined leaves and, even, branched 'flowering' stems, all springing from a central point. Move the seedlings on a damp day as soon as recognized, and take care with the long tap-root. (It cannot be moved in maturity and survive.) Allow plenty of space between each plant and plant it against a brick wall, or near an acid-green, red, or grey-leaved plant, where the contrasting form and leaf-coloration of the Caper Spurge will show to the best advantage.

CONTROL Easily pulled up in infancy but wear gloves, because the latex in the plant can burn the skin. Euphorbias are very good in the compost-heap.

EUPHORBIACEAE Spurge Family

Petty Spurge
Euphorbia peplus L.

A Small and bushy
AH 10–30 cm (4–12 in)
L Alternate, obovate, obtuse
R Slender and vertical
F Very small, 'flowers' are bracts. April to November
H Cultivated and waste land
Annual
Poisonous

When picked, this plant, like all the typical Euphorbias, yields an acrid, milky latex which coagulates on drying. Its flowers are too minute to be noticed but it has enlarged bracts like green petals which can look very attractive.

CONTROL Easily weeded out, but wear gloves to do so. The 'milk' from the stems may burn the skin, and affect the eyes.

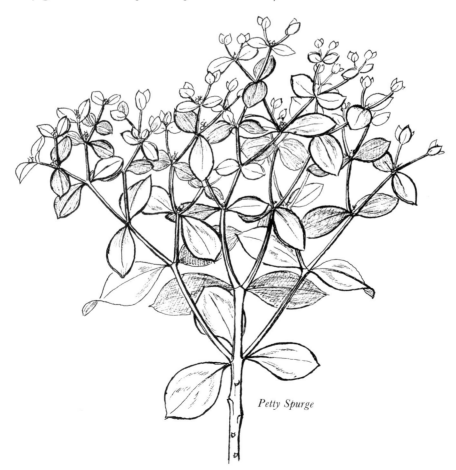

Petty Spurge

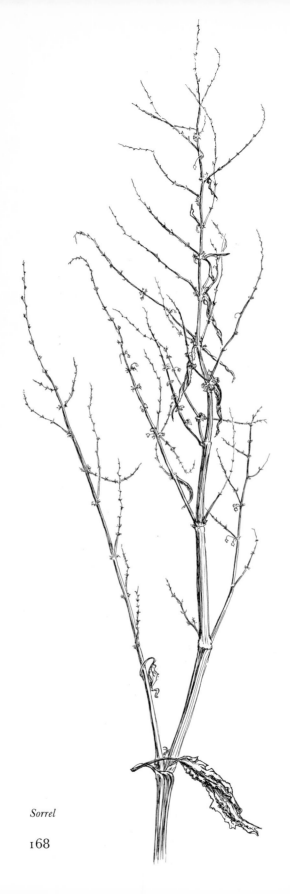

Sorrel

POLYGONACEAE Dock Family

Sorrel
Rumex acetosa L.

A   Tall and vigorous
AH   Up to 100 cm (40 in)
L   Oblong-lanceolate, hastate
R   Long
F   Leafless and in a spike. May to June
H   Grassland
     Perennial

Sorrel is carefully avoided by grazing animals, so all the seeds are free to fall to produce yet more plants, which makes this weed very common in pasture.

USES It is good for some ailments, though not for gout or rheumatism. A decoction, made as follows, is remedial in cases of acne and eczema: 1 oz (30 g) leaves to 2 pt (1.14 l) boiling water, allow to infuse, strain. Dose: 1 cupful 3 times a day. For boils, carbuncles, abcesses and similar eruptions, cook the Sorrel leaves and apply them, warm, as a poultice to the affected area, which will bring the pustule more quickly to a head.

For polishing and preserving bamboo and wicker furniture or for an extra shine to silver plate, rub over with a cloth wrung out in water in which Sorrel leaves have been boiled; polish after with a clean dry cloth.

Ink-stains (not ball-point ink) on white material may be removed by rubbing the mark with some Sorrel leaves, then rubbing with soap; rinse well and repeat as needed.

In France, Sorrel can be purchased in the street markets and is often added to thick soups made of haricot beans, lentils and potatoes. Sorrel soup is made as follows:

Take 1 lb (455 g) fresh leaves, an onion, Rosemary, 1 oz (30 g) flour, 3 oz (85 g) butter, 2 oz (55 g) white breadcrumbs, 2 eggs, ¾ pt (0.4 l) cream and 4 pt (2.30 l) water. Chop or tear the leaves: melt the butter in a heavy pan and add the leaves, chopped onion, Rosemary and flour. Put the water on to heat and, while it is coming to the boil, add the breadcrumbs and seasoning to the mixture. Stir well all the time, and add the boiling water slowly. Cook over a low heat for an hour. Make a mix of the 2 well-stirred egg-yolks and cream, and add to the soup just before serving.

CONTROL The same as for *Rumex obtusifolius* (below).

POLYGONACEAE Dock Family

Broad-leaved Dock, Docken

Rumex obtusifolius L.

A   Erect and branched
AH 50–100 cm (20–40 in)
L   Ovate-oblong, base cordate, margins undulate
R   Long
F   Branched, leafy, with distinct whorls. June to October
H   Waste ground, hedgerows, field-edges
    Perennial

This dock, which can be rubbed on nettle stings, always seems to grow conveniently near to the nettle-beds and is a familiar sight in the autumn and winter with its dead spires of rusty seeds. It is the food plant of one butterfly, the Small Copper, *Lycaena phlaeas* L. and various moths, such as the Setaceous Hebrew Character moth, *Amathes c-nigrum* L., the Bird's Wing moth, *Dypterygia scabriuscula* L. and the Clouded Buff moth, *Diacrisia sannio* L.

CONTROL Docks have long, strong and deep-penetrating tap-roots, which are impossible to get out from among the roots of the rightful inhabitants of a rockery. Either keep cutting off all the aerial parts until the dock finally gives up, or cut off the top growth and touch the cut area with a paint-brush dipped in a spot-treatment weedkiller. For other areas of the garden, general deep digging is the only natural method of prevention, though each piece of snapped-off root must be collected and burnt.

Broad-leaved Dock

URTICACEAE Nettle Family

Stinging Nettle
Urtica dioica L.

A   Tall and vigorous
AH 30–150 cm (12–60 in)
L   Hispid, ovate, acuminate, dentate, alternately
    paired
R   Much branched, tough, yellow and creeping
F   Axillary, long, green and catkin-like. May to
    October
H   Common everywhere
    Perennial

Culpeper said that Stinging Nettles need no description 'as they may be found by feeling in the darkest night'. The stinging hairs are hollow and are attached to a cushion of delicate cells which, when the hair is touched or brushed the wrong way, emit a juice chiefly composed of formic acid and ammonia. It is a strange fact that the juice of the plant is an antidote to its own sting, and dock leaves, which are usually to be found growing nearby, contain the same soothing substance. Being of the same family, Rhubarb leaves will also soothe the blistered area, as will crushed Rosemary, Mint and Sage.

The Stinging Nettle grows throughout the temperate regions of the world, and is to be found in all the waste corners of civilization in Australia, the Americas, South Africa and Japan. If a garden is full of Stinging Nettles, the soil is rich in nitrogen. The plant is host to the spiky black caterpillars of the Peacock butterfly (*Nymphalis io*), and apples are said to ripen more quickly and grow bigger if they dangle over a nettle bed.

Stinging Nettle

USES Nettles can be cooked and eaten, made into a pleasant beer, used as fodder for cattle, food for poultry and fibre for woven cloth. A safe, home-made pesticide may be made by boiling the roots and leaves in water for half an hour, straining, cooling, and spraying on aphides and flea-beetles. Nettles make wonderful compost, as they stimulate fermentation in the heap, and, when hung in the larder together with Tansy (*Tanacetum vulgare*), they keep the flies away.

Only cook nettles in the spring. Pick the young tops, wearing gloves, wash in cold running water and put in a saucepan with no more water added. Cook as for spinach with the lid on for about 20 minutes. Chop them, run through a sieve, add pepper, salt and butter and serve with poached eggs. This dish of greens has a slightly laxative effect.

In bad cases of nosebleeding, place a small piece of gauze, well-saturated with nettle juice, into the nostril. An infusion of the fresh leaves is soothing for burns and accelerates healing, and the dried leaves, burned and inhaled, relieve bronchial and asthma sufferers.

To stimulate hair growth, comb the hair daily with fresh nettle juice.

CONTROL The best method is to cut them down, and compost them, and then to rotovate. If nettles are cut down three times a year for three years in succession, they will probably give up the struggle to survive.

COMPOSITAE Daisy Family

Pineapple Weed
Matricaria matricarioides (Less.) Porter

A   Inconspicuous with feathery leaves
AH 5–30 cm (2–12 in)
L   Glabrous and much-branched, 2–3 times
    pinnate, linear, bristle-pointed
F   Aromatic, no ray-florets, disk-florets a
    greenish-yellow cone. June to July.
H   Waysides, tracks and gateways
    Annual

This daisy of the cart-rut has no discernible petals, but nevertheless has the strong sweet scent of fresh pineapple. It came originally from the inhospitable wastes of north-east Asia and its spread has been helped by the advent of the motor-car and the subsequent farm-tractor – the rough outer-surface of the seeds helps them to stick in the mud of car and tractor tyres.

It does no harm if confined to a contrived cranny in

Pineapple Weed

the path, where the delicious scent of the trodden leaves and flower-heads will scent the air on a damp day.

CONTROL It is very easily removed by hand and contains minerals such as calcium which are most helpful in the compost-heap.

JUNCACEAE Rush Family

Soft Rush
Juncus effusis L.

A Stiffly erect and tufted
AH 30–150 cm (12–60 in)
L None
F In a tuft, from below top of stem
H Wet pastures and damp woods
Perennial

If you have the porcupine spikes of Soft Rush in your lawn, it will need proper draining. Mowing will certainly hide the evidence temporarily, but the job of laying drains to carry away the excess water will have to be faced sooner or later.

DIOSCOREACEAE Yam Family

Black Bryony, Beadbine
Tamus communis L.

A Climbing
L Broadly ovate, deeply cordate at base, acuminate, dark shining green
R Large blackish tuber
F Axillary, in racemes, dioecious, yellowish-green. May to July
H Wood-margins, hedgerows, scrub
Perennial
Poisonous

The twisted scarlet 'bead-necklaces' of Black Bryony in autumn are often the first signs of its existence to be noticed. All parts of the plant are poisonous, the root being the most toxic. It was formerly used medicinally but is now considered too dangerous.

Black Bryony is not related to White Bryony (*Bryonia dioica*) and is dissimilar in its leaf shape, though both have red berries in autumn. The word 'bryony' is derived from a Greek word meaning to grow rapidly, which could equally well be given to many of our climbing plants.

Black Bryony

Soft Rush

CONTROL This very poisonous plant must be treated with care. Wear gloves if handling the long, vine-like ropes of leaves, and especially when digging up the black-skinned root – if it can be got at. If it cannot, keep cutting off all the growing shoots until it is eventually discouraged.

ARACEAE Arum Family

Lords and Ladies, Cuckoo-pint, Jack-in-the-Pulpit, Wake-Robin, Friar's Cowl, Wild Arum

Arum maculatum L.

A   Seasonal leaves and erect flowering spathe
AH 30–50 cm (12–20 in)
L   In spring, triangular hastate, often purple-spotted
R   Tuber
F   Pale yellow-green spathe, sometimes edged and spotted purple. Purple spadix. April to May
H   Woods and shaded hedgebanks
    Perennial
    Poisonous

Lords and Ladies

173

Annual Meadow Grass

Rye Grass

Arum maculatum has over seventy local names and has a powerful reputation as a magical plant. Because of its physical shape, it was used in many spells to improve sexual potency or, possibly, to retard it. One spell instructs as follows: take earth or dough and make a doll to represent the man, naked and with his organ clearly defined and erect. Baptize the doll in the name of the person, bury it in the ground and plant Wild Arum over it. This should be done during the waxing moon and in secret. As the Wild Arum grows the man's sexual abilities grow; if the Wild Arum dies, he will become impotent.

And another spell – to retard ageing lines: pick the Cuckoo-pint while the dew is still on it and wash your face in the dew and flowers. Do this sequentially for seven days during the period of the waxing moon, at dawn and by running water.

The roots or tubers contain a high proportion of starch which was extracted, dried and made into a food called Portland sago. (The toxins present in the plant are destroyed in the action of drying.) In medieval days, starch was made from the roots, but as Gerard says, this was 'most hurtfull for the hands of the laundresse that have the handling of it, for it chappeth, blistereth, and maketh the hands rough and rugged and withall smarting'.

CONTROL Constantly root up the tubers of the plant, which does not like to be disturbed.

GRAMINEAE Grass Family

Rye Grass, Tinker-tailor Grass
Lolium perenne L.

A   Wiry with flattened flowering head
AH 25–50 cm (10–20 in)
L   Smooth, folded when young
F   Spikelets of 3–11 florets, variable. May to August
H   Waste places
    Perennial

A brushy grass that furs road verges, *Lolium perenne* seems to prefer dusty, barren patches of ground and if the soil is improved by fertilizer, it will disappear.

CONTROL Deep digging of the area to remove the rooted plants before seeding.

GRAMINEAE Grass Family

Annual Meadow Grass
Poa annua L.

A   Tufted and feathery
AH 5–30 cm (2–12 in)
L   Flat, slightly keeled, smooth
F   Panicle greenish-white or purplish, triangular.
    Throughout the year
H   Found everywhere
    Annual

A meadow of this very common variety of grass, with its feathery flower panicles shimmering in waves of shaded colour, is balm to the soul.

CONTROL If you have a large patch of it unmixed with anything else, rotovate if there is space; otherwise dig over.

GRAMINEAE Grass Family

Cock's-foot
Dactylis glomerata L.

A   Coarse and typical
AH Up to 1 m (3 ft)
L   Flat, rough, keeled
F   Horizontally branched panicle, crowded at the
    ends of the spikelets, green and violet. May to
    August
H   Found everywhere
    Perennial

So-called because of the flowering head which resembles an outstretched cock's foot, this grass is quite distinct in its appearance from any other. It is the food plant of the Wall Brown butterfly, *Pararge megaera* L.

CONTROL Thorough digging.

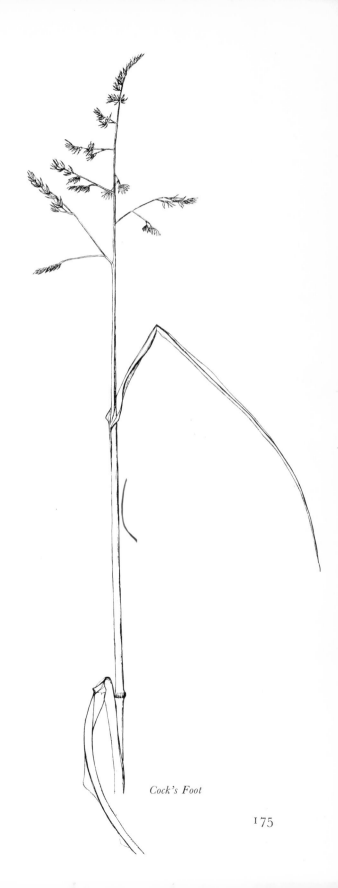

Cock's Foot

175

Barren Brome

Barren Brome
Bromus sterilis L.

A Curving stems with swinging flower-heads
AH 30–100 cm (12–40 in)
L Soft, flat and downy
F Drooping panicle. May to July
H Waste places, roadsides
    Annual

The swinging seed-heads of this grass are particularly pleasing when dried and added to thicker, dried material, but do not hang this grass upside down or it will lose its graceful curve. The rough 'awns' are collected by passing animals and so dispersed.

CONTROL Thorough digging over.

GRAMINEAE Grass Family

Couch Grass, Twitch, Scutch, Quickgrass, Quicken
Agropyron repens (L.) Beauv.

A Tall and inconspicuous
AH 30–100 cm (12–40 in)
L Narrow
R Rhizomes abundant and far-creeping
F Spike with offset glumes, variable. June to September
H Fields, waste places and gardens
    Perennial

The one grass name familiar to all gardeners, as is its physical appearance, it is impossible to ignore the persistent Couch Grass once you have it. It will grow through any obstruction, such as potatoes, parsnips, walls and the bottom of your neighbour's fence.

However, it has a magical reputation and may always be useful as a spell 'to entrance your hearers and make them believe you': go, just before dawn, to where the Couch grows, pull up some of its roots saying 'As the birds wake and sing, so will I wake and sing; as we listen to them so they listen to me'. Then wash the roots in running water and eat them. This should be done three mornings in succession during the waxing moon.

USES For centuries Couch Grass has been used as a remedy for bladder and kidney disorders, cystitis, renal colic, jaundice and inflammation of the bowel, and a course of the following decoction is said to bring considerable relief from these maladies.

Take 1 oz (30 g) root or rhizome and soak in cold water for 3 hours. Bring to the boil and boil for 1 minute. Throw this water away. Crush the roots and return them to 2 pt (1.14 l) fresh hot water. Reduce by boiling to 1 pt (0.6 l) and allow to cool. Dose: 3 cupfuls a day. The flavour is strong, and may be disguised by adding a little liquorice when the roots are almost finished boiling, or by adding slices of lemon or orange and honey to taste.

This is the grass that cats and dogs eat to keep themselves healthy.

CONTROL Pull it up, hand-over-hand. Even though its roots may go down very deep, destruction of part of the growth will stop regeneration for a year or two. Chemical control is only partly effective, as the really deep-seated roots are not reached by herbicides. If you compost it, note that the roots need a lot of heat to rot them, and if they are not entirely decomposed, you will merely be replanting your own Couch.

GRAMINEAE Grass Family

Wall Barley, Wild Barley
Hordeum murinum L.

A   Stout
AH 20–60 cm (8–24 in)
L   Flat and pilose on both sides
F   Spikelets with all glumes awned. June to July
H   Waste places
    Annual

This wild barley was cultivated as a food plant by primitive peoples, and 'kernels' of Wall Barley, among many other wild plants, were found to be part of the primitive gruel which was Tollund man's last meal. (He died in Denmark about 2,000 years ago.)

CONTROL Dig over the area thoroughly before the seeds form.

Wall Barley

Couch Grass

177

Oat-grass   *Common Bent*   *Timothy*

GRAMINEAE Grass Family

Oat-grass, False Oat

Arrhenatherum elatius (L.) Beauv.

A   Tall and whiskery
AH 60–120 cm (24–48 in)
L   Flat, scabrid
F   Panicle lax, nodding, narrow. June to August
H   Rough grassy places
    Perennial

This tall, rather hairy-looking grass can spring up anywhere, particularly after delivery of a load of farmyard manure.

CONTROL Dig deep enough to remove the roots.

GRAMINEAE Grass Family

Common Bent, Bent-grass

Agrostis tenuis Sibth.

A   Tall and delicate
AH 2–100 cm (1–40 in)
L   Flat, slightly scabrid
R   Rhizomatous
F   Panicle effuse, ovoid, cylindrical or pyramidal.
    June to August
H   Acid grassland
    Perennial

Gracefully shaped, this grass can be dried and used in flower arrangements, though it is very fragile when dry.

CONTROL Dig deep enough to remove the roots.

GRAMINEAE Grass Family

Timothy

Phleum pratense L.

A   Neat and cylindrical flower-heads
AH 50–100 cm (20–40 in)
L   Very rough
R   Stem sometimes tuberous at base
F   Cylindrical panicle. July
H   Meadows
    Perennial

This neat-topped grass is easily recognized – it looks as symmetrical as a miniature bullrush. It is often sown for hay grass and is the food plant of the Marbled White butterfly, *Melanargia galathea* L.

CONTROL Dig deep enough to remove the roots.

Bibliography

Boland, Maureen and Bridget, *Old Wives' Lore for Gardeners* (Bodley Head, 1976)

Carey, M. C., *Flower Legends* (C. Arthur Pearson, 1929)

Clapham, Tutin and Warburg, *Flora of the British Isles* (Cambridge University Press, 1962)

Clapham, Tutin and Warburg, *Flora of the British Isles; Illustrations*, Volumes 1–4 (Cambridge University Press, 1963)

Connell, Charles, *Aphrodisiacs in your Garden* (Arthur Barker, 1965)

Courtenay, B. and Zimmerman, J., *Wild Flowers and Weeds* (Van Nostrand Reinhold, 1973)

Culpeper, Nicholas, *Culpeper's Complete Herbal* (1653)

Culpeper, Nicholas, *Complete Herbal* (W. Foulsham, 1952)

Fernie, W. T., *Herbal Simples* (John Wright, 1914)

Flück, Hans, translated by J. M. Rowson, *Medicinal Plants* (Foulsham, 1976)

Genders, Roy, *The Scented Wild Flowers of Britain* (Collins, 1971)

Gerard, John, *Herball* (Minerva, 1971)

Grieve, Maude, *A Modern Herbal* (Jonathan Cape, 1974)

Griffith-Jones, Joy, *The Value of Weeds* (Soil Association, 1975)

Grigson, Geoffrey, *A Herbal of All Sorts* (Phoenix House, 1959)

Grigson, Geoffrey, *The Englishman's Flora* (Paladin, 1975)

Harris, Ben Charles, *Eat the Weeds* (Keats Pub., 1975)

Hatfield, Audrey Wynne, *How to Enjoy Your Weeds* (Muller, 1974)

Hill, Sir John, *The Family Herbal* (C. Brightly, 1812)

Horwood, A. R., *British Wild Flowers in their Natural Haunts* (Gresham Pub. Co., 1919)

House, Homer D., *Wild Flowers* (Collier-Macmillan, 1975)

Jordan, Michael, *A Guide to Wild Plants* (Millington, 1976)

Kadans, Joseph, *Encyclopedia of Medicinal Herbs* (Arco, 1975)

Levy, Juliette de Bairacli, *The Illustrated Herbal Handbook* (Faber, 1974)

Lust, John, *The Herb Book* (Bantam, 1975)

Mabey, Richard, *Food for Free* (Collins, 1972)

Martin, W. Keble, *Concise British Flora in Colour* (Michael Joseph and Ebury Press, 1969)

McClintock, David and Fitter, R. S. R., *Wild Flowers* (Collins, 1956)

Morley, Brian D., *Wild Flowers of the World* (Octopus, 1974)

Palaiseul, Jean, *Grandmother's Secrets* (Barrie and Jenkins, 1973)

Perry, Frances, *Flowers of the World* (Hamlyn, 1972)

Philbrick, H. and Gregg, R., *Companion Plants* (Robinson and Watkins, 1967)

Polunin, Oleg and Smythies, B. E., *Flowers of South-West Europe – A Field Guide* (Oxford University Press, 1973)

Pratt, Anne, *Wild Flowers of the Year* (Lutterworth, 1940)

Salisbury, Sir Edward, *The Living Garden* (G. Bell, 1946)

Skinner, Charles M., *Myths and Legends of Flowers, Trees, Fruits and Plants* (Lippincott, 1911)

Step, Edward, *Herbs of Healing* (Hutchinson, 1926)

Weeks, Nora and Bullen, V., *The Bach Flower Remedies* (C. W. Daniel, 1976)

Whitehead, Stanley B., *Garden Weeds and their Control* (Dent, 1949)

Guide to Chemical Control of Weeds

Chemical herbicides, or weedkillers as they are better known, have been in general use only since the beginning of this century, and for the first half of that time were confined to inorganic chemicals, that is, those not containing carbon compounds in their constituents. In the 1930s and 1940s, the first organic weedkillers were introduced, to be followed by those directly related to the hormone structures of specific plants. It was these, and the more-and-more complex formulae that were developed from them, that have led to the great range of total and selective weedkillers that we can buy today.

All weedkillers, total or selective, kill by disturbing one or more aspects of the plant's anatomy, metabolism and physiology, or the delicate balance between them. Sodium chlorate, for instance, a well-known total killer, disturbs the action of the cells that convert nitrates in the presence of light – a process essential to all plant life – so that all plants suffer accordingly. Selective weedkillers, on the other hand, are chemically structured to attack the specific metabolism of the plant to be destroyed, without harming nearby plants that might take in the herbicide. Chlorpropham, for instance, attacks the respiration of a wide range of germinating weeds in fruit beds without effect on the fruits themselves, while 2,4–D, one of the phenoxy acids, affects the chromosomes and enzymes of specific lawn weeds, leading to cell distortion, chemical changes and death.

There are three basic types of herbicide.

1. Foliage-applied, contact herbicides

These kill only the foliage that they touch, either from a spray, or a dribble bar or, in individual 'spot' treatment, from an aerosol, brush or rag. Some are selective, such as Ioxynil, which kills weeds in newly sown lawns, and others are non-selective, such as paraquat-diquat mixtures, which kill all green matter that they touch. These herbicides have no effect on the roots of perennial weeds.

2. Foliage-applied, translocating herbicides

Again applied by spray or other wetting means, these herbicides enter the plant through its foliage and 'translocate' within it, moving throughout its structure, down to the roots. Usually it takes 2 to 3 weeks to complete the translocation and kill the weed, which often twists before dying owing to internal cell distortion. As with contact herbicides, some are selective, such as MCPA and Mecoprop, which are used to kill broad-leaved weeds in established lawns, and they can be used among ornamental plants provided that they do not come into contact with them. Further, the spray must be kept off the soil, because wanted plants can take up the chemicals through their roots. Others are non-selective, such as Sodium chlorate, which kills everything, including deep-rooted perennial weeds.

Some herbicides of this type are both selective and non-selective, depending on the dose rate. Dalapon, for instance, is selective at low rates, and can be used among fruit and vegetables to suppress couch grass. At high dose rates, however, it is non-selective and will kill all grasses and grass-type weeds.

3. Soil-applied, residual herbicides

These are taken up by the weeds through their roots, and when applied to weed-free soil, they kill the weeds as they germinate. One example, Propachlor, is usually effective for several weeks, while Simazine remains effective for several months, but it cannot be used in vegetable plots. As with some of the translocating herbicides, some of the soil-acting chemicals may be selective or non-selective, according to dose rate. For example, a low dose rate of Simazine may be used selectively in rose-beds, whereas a high dose rate can be used on paths to kill almost every plant.

The way in which a selective herbicide can be turned into a devastating exterminator simply by changing its se rate is one good reason for following manu-facturers' instructions exactly, no matter how stringent they appear to be.

The second reason for following instructions meticulously is of greater consequence. All herbicides are dangerous to differing degrees. They can kill not only plant life but animal life too if used improperly, and animal life includes you. Spray on a windy day, and you might well damage, irreparably, your neighbour's roses, or kill the fish in your pond, or find, if you were not wearing goggles, that your eyelids feel like sandpaper. Let some of the organic chemicals into running water, and you may kill everything down-stream, both in the water and where it is used for irrigation or drinking. That, certainly, is a statutory offence. Store partly used mixtures in old soft-drink bottles and your children might die.

Keep partly used containers tightly shut, per-manently labelled, out of reach of children and ani-mals, and away from food. That is common sense. Read the instructions, and then read them again.

The following table shows which chemicals may be used safely, and for what purpose, in different parts of the garden.

HERBICIDES TO USE IN DIFFERENT PARTS OF THE GARDEN

Parts of the garden	Weeds to be controlled	Chemicals to use	Notes
Established lawns	Annual and perennial broad-leaved weeds	Ioxynil 2,4–D MCPA Mecoprop Dichlorprop 2,4–D/Mecoprop mixtures 2,4–D/Dichlorprop mixtures	Apply in good growing conditions when rain is not expected for several hours. Several applications might be needed, 4 to 6 weeks apart.
	Moss	Ferrous sulphate Mercurous chloride Chloroxuron Dichlorophen	Temporary control only. Check aeration and drainage of lawn.
Newly sown lawns	Broad-leaved weeds	Ioxynil	Do not apply for at least 6 months after sowing. Chemicals suitable for established lawns should not be used.
Weed-free paths and drives		Simazine	Apply in the spring to prevent weed germination for the whole season. Avoid underlying roots of trees and shrubs.
Weedy paths and drives	Annual and perennial weeds	Simazine at high dose rate Simazine/MCPA mixtures Simazine/Aminotriazole mix	
	Annual weeds	Paraquat/Diquat	
	Couch grass	Dalapon	
	Obstinate weeds	2,4–D MCPA Dichlobenil	

Parts of the garden	Weeds to be controlled	Chemicals to use	Notes
Neglected paths and drives	Established weeds	Sodium chlorate	
Fruit bushes and trees	Weed-free	Simazine	As for weed-free paths.
	Established annual weeds	Paraquat/Diquat	
	Established perennial weeds	Dichlobenil	Apply only after fruit bushes are 2 years old. Use in late winter/early spring.
	Couch grass	Dichlobenil Dalapon	Apply in November.
Rose-beds	Weed-free	Simazine	As for weed-free paths, but at lower rates. Controls annual weeds for the season. Not suitable for beds containing bulbs and bedding plants.
	Annual and perennial weeds mixed with bulbs and bedding plants	Propachlor Chloroxuron Paraquat/Diquat Dichlobenil	
Shrubbery	General weeds with no herbaceous plants	Simazine Dichlobenil	
	General weeds with herbaceous plants	Propachlor Chloroxuron	
Vegetable and flower-beds	Weed-free	Propachlor Chloroxuron	Selectively controls germinating annual weeds for 6 to 8 weeks among established plants. Will not control established weeds. Chloroxuron should not be used before the vegetable seedlings possess 5 true leaves.
	General weeds	Paraquat/Diquat	Use with extreme care to avoid contact with wanted plants. Inactivated on contact with the soil. Seeds may be sown immediately after application and plants put in after 24 hours.
Neglected or new areas to be cultivated	General weeds	Paraquat/Diquat	For dessication prior to burning and digging. Crops may be planted immediately.
		Sodium chlorate	Do not plant for 12 months after treatment.
	Grass weeds	Dalapon	Do not plant for 6–8 weeks after treatment.
	Nettles, docks and thistles	2,4-D 2,4,5-T MCPA 2,4,D/2,4,5-T mixtures	

The information contained in the above table is based on the 1977 *Directory of Garden Chemicals* published by the British Agrochemicals Association and on the 1976 edition of *Chemical Weed Control in your Garden* published by the Agricultural Research Council Weed Research Organisation.

Index of Common Names

Index of Scientific Names

Acer pseudoplatanus, 160, *161*
Achillea millefolium, 94, *95*
Aconitum anglicum, *128*, 129
Aegopodium podagraria, 75, *76*
Aethusa cynapium, 77, *77*
Agrimonia eupatoria, 35, *36*
Agropyron repens, 176, *177*
Agrostis tenuis, *178*, 179
Alliaria petiolata, 65, *65*
Allium ursinum, 98, *98*
Anagallis arvensis, 118, *121*
Anthriscus sylvestris, 74, *75*
Aphanes arvensis, 161, *162*
Aquilegia vulgaris, 136, *136*
Arctium lappa, 150, *151*
Arrhenatherum elatius, *178*, 179
Artemisia vulgaris, 50, *51*
Arum maculatum, 173, *173*
Atropa bella-donna, 143, *143*

Barbarea vulgaris, 22, *22*
Bellis perennis, 94, *96*
Bromus sterilis, 176, *176*
Bryonia dioica, 163, *163*
Buddleja davidii, 138, *141*

Calendula officinalis, 58, *59*
Calluna vulgaris, 116, *116*
Calystegia sepium, 82, *82*
Capsella bursa-pastoris, 62, *63*
Cardamine hirsuta, 64, *64*
Centranthus ruber, 122, *124*
Cerastium holosteoides, 65, *66*
Chelidonium majus, 18, *20*
Chenopodium album, 159, *160*
Chenopodium polyspermum, *159*, 159
Chrysanthemum parthenium, 97, *97*
Chrysanthemum segatum, *49*, 50
Circaea lutetiana, 114, *114*
Cirsium arvense, *152*, 153
Cirsium vulgare, 151, *152*
Convolvulus arvensis, *118*, 119
Conyza canadensis, 93, *93*
Corydalis lutea, 19, *21*

Cymbalaria muralis, 146, *146*

Dactylis glomerata, 175, *175*
Datura stramonium, *84*, 86
Digitalis purpurea, 119, *119*
Dipsacus fullonum, 149, *150*
Dryopteris dilatata, 158, *158*

Endymion non-scriptus, 131, *133*
Epilobium angustifolium, 112, *113*
Epilobium montanum, 112, *113*
Equisetum arvense, *156*, 157
Erica cinerea, *116*, 117
Erica tetralix, 116, *117*
Euphorbia lathyrus, 166, *166*
Euphorbia peplus, 167, *167*

Foeniculum vulgare, 39, *41*
Fragraria vesca, 71, *72*
Fumaria officinalis, 102, *104*

Galium aparine, *90*, 91
Galium verum, *45*, 46
Geranium dissectum, 103, *104*
Geranium molle, 103, *105*
Geranium robertianum, *105*, 106
Geum urbanum, 34, *36*
Glechoma hederacea, 149, *149*

Hedera helix, 162, *164*
Helxine soleirolii, 115, *115*
Heracleum mantegazzianum, 79, *80*
Heracleum sphondylium, *78*, 79
Hesperis matronalis, 137, *137*
Hordium murinum, 177, *177*
Hypochoeris radicata, 51, *52*

Impatiens glandulifera, 106, *108*
Iris foetidissima, 153, *153*

Juncus effusis, 172, *172*

Lamium album, *85*, 86
Lamium purpureum, 122, *124*

Lapsana communis, 51, *52*
Ligustrum vulgare, *81*, 82
Linaria Cymbalaria, 146, *146*
Linaria purpurea, 144, *145*
Lolium perenne, 174, *174*
Lotus corniculatus, 31, *32*
Lunaria annua, 136, *137*
Lupinus arboreus, *26*, 27
Lysimachia nummularia, 43, *44*

Malva sylvestris, 102, *104*
Matricaria inodora, 96, *96*
Matricaria matricarioides, 171, *171*
Medicago arabica, 29, *29*
Medicago lupulina, 27, *28*
Melilotis altissima, 30, *30*
Mentha spicata, 147, *148*
Mercurialis perennis, *165*, 165–6
Montia perfoliata, 70, *70*
Myosotis sylvatica, 129, *129*

Odontites verna, 120, *120*
Oenothera biennis, 38
Oenothera erythrosepala, 37, *39*
Oxalis corniculata, 23, *25*

Papaver rhoeas, *100*, 101
Papaver somniferum, 101, *102*
Pastinaca sativa, 42, *43*
Petasites fragrans, 123, *125*
Phleum pratense, *178*, 179
Picris echioides, 53, *53*
Plantago lanceolata, 88, *89*
Plantago major, 87, *87*
Poa annua, *174*, 175
Polygonum aviculare, 114, *115*
Polygonum cuspidatum, 79, *81*
Polygonum persicaria, 115, *116*
Potentilla anserina, 31, *33*
Potentilla reptans, *33*, 34
Prunella vulgaris, 148, *148*
Pteridium acquilinum, *157*, 157

Ranunculus acris, 16, *16*
Ranunculus bulbosus, *17*, 18
Ranunculus ficaria, 18, *20*

Ranunculus repens, 16, *17*
Rubus ulmifolius, *110*, 111, *111*
Rumex acetosa, 168, *168*
Rumex obtusifolius, 169, *169*

Sagina procumbens, 69, *69*
Sambucus nigra, 91, *92*
Sedum acre, 37, *38*
Sedum album, 73, *73*
Senecio jacobaea, *45*, 46
Senecio vulgaris, 47, *47*
Sinapsis arvensis, 19, *21*
Sisymbrium officinale, 22, *24*
Smyrnium olusatrum, 39, *40*
Solanum dulcamara, 144, *145*
Solanum nigrum, 83, *83*
Sonchus asper, *56*, 57
Sonchus oleraceus, 54, *55*
Stachys sylvatica, *121*, 122
Stellaria holostea, 68, *68*
Stellaria media, 66, *67*
Symphytum officinale, 139, *142*

Tamus communis, 172, *172*
Taraxacum officinale, 57, *57*
Thlaspi arvense, 62, *62*
Tragopogon pratensis, 54, *54*
Trifolium arvense, 71, *71*
Trifolium dubium, 31, *32*
Trifolium hybridum, 107, *107*
Trifolium pratense, 107, *109*
Trifolium repens, 70, *72*
Tropaeolum majus, 23, *25*
Tussilago farfara, 48, *48*

Ulex europaeus, 27, *28*
Umbilicus rupestris, 74, *74*
Urtica dioica, 170, *170*

Valerianella locusta, 130, *131*
Verbascum thapsus, 44, *44*
Veronica chamaedrys, 130, *132*
Veronica persica, 130, *132*
Vicia cracca, 138, *140*
Vicia sativa, 109, *109*
Vinca minor, 139, *141*
Viola tricolor, 138, *140*